MW01070495

Cooperative Chemistry

Laboratory Manual

Fifth Edition

Melanie M. Cooper

Clemson University

Mc Graw Hill

Connect Learn Succeed™

The McGraw-Hill Companies

COOPERATIVE CHEMISTRY LABORATORY MANUAL, FIFTH EDITION

Published by McGraw-Hill, a business unit of The McGraw-Hill Companies, Inc., 1221 Avenue of the Americas, New York, NY 10020.

This book is printed on recycled, acid-free paper containing 10% postconsumer waste.

Printed in the United States of America.

3 4 5 6 7 8 9 0 QDB / QDB 1 0 9 8 7 6 5 4 3 2

ISBN 978-0-07-340272-7
MHID 0-07-340272-9

Vice President & Editor-in-Chief: *Marty Lange*
Vice President EDP/Central Publishing Services: *Kimberly Meriwether David*
Publisher: *Ryan Blankenship*
Executive Editor: *Jeff Huettman*
Executive Marketing Manager: *Tamara L. Hodge*
Development Editor: *Shirley Oberbroeckling*
Senior Project Manager: *Joyce Watters*
Design Coordinator: *Brenda Rolwes*
Cover Designer: *Studio Montage, St. Louis, Missouri*
Cover Image: *Courtesy of the author*
Buyer: *Susan K. Culbertson*
Compositor: *Aptara, Inc.*
Typeface: *12/14 Times Roman*
Printer: *QuadGraphics*

Some of the laboratory experiments included in this text may be hazardous if materials are handled improperly or if procedures are conducted incorrectly. Safety precautions are necessary when you are working with chemicals, glass test tubes, hot water baths, sharp instruments, and the like, or for any procedures that generally require caution. Your school may have set regulations regarding safety procedures that your instructor will explain to you. Should you have any problems with materials or procedures, please ask your instructor for help.

www.mhhe.com

About the Author

Melanie Cooper received her Ph.D. in chemistry from the University of Manchester in England. She is at present an Alumni Distinguished Professor of Chemistry at Clemson University in South Carolina. Dr. Cooper's research interests lie in the area of Chemistry Education, including research into problem solving in chemistry, laboratory development, and the development of active learning methods for large enrollment courses. She has been the recipient of a number of grants and awards in this area.

This lab manual is based on work that was funded by the National Science Foundation (Grant#9155954) and the Fund for the Improvement of Postsecondary Education (FIPSE Grant#P116B20009-92A).

ACKNOWLEDGMENTS

I would like to acknowledge Barbara Lewis, Will Wannamaker, and all the undergraduate students who have taken general chemistry laboratories at Clemson, and the graduate teaching assistants who have all provided valuable feedback on their experiences.

Contents

Section 1

Cooperative Chemistry:

How and Why

COOPERATIVE CHEMISTRY LABORATORIES

To the Instructor

This lab manual is the product of over fifteen years of laboratory development at Clemson University and was initially funded by the National Science Foundation. The laboratory program began to evolve after the author realized the more traditional "cookbook" type laboratories prevalent in many general chemistry courses were squandering the opportunity to excite students and show them how real science is done. The laboratory program described in this manual attempts to give students an idea of the processes that scientists go through as they develop their ideas and design new experiments. Most scientists would agree that this is a laudable goal. But there are many reasons why general chemistry labs have degenerated into the rather simplistic exercises that can be graded based on the students' proximity to the "correct answer." The logistics of running a more complex program, in which the students have more input and the grading must be somewhat more subjective, are rather daunting. Both this manual and the accompanying instructor's manual attempt to present ways in which a course like the one at Clemson can be adopted and adapted for use at other institutions.

Some obviously will ask if changing the nature of the labs is worth the effort. Judging from the results we have observed, the answer is an emphatic yes. We have found that students respond very well to laboratories using this format. When traditional laboratories and cooperative laboratories were run side by side and students were arbitrarily assigned to either type of lab, students in the cooperative labs had significantly more positive attitudes toward science and the laboratory experience; they also thought they had learned more. In addition we found that female students' retention rates in cooperative labs were higher than their peers in regular labs. Female students in cooperative labs also tended to score better than their peers on the examinations in the lecture portion of the course.

The projects in this manual can be used in a variety of contexts and with a variety of different students. Although most of the laboratory curriculum development has been in the science and engineering major's course, we have, for example, used some of the projects (namely concrete and the electrochemistry project) for nonscience majors. The outcomes and expectations for each project will depend on the background of the students, but it seems to be true that all types of students can benefit from this type of laboratory experience.

Before embarking on changes such as those presented in this lab manual, instructors should give some thought to what they believe to be the purpose of the laboratory. In the Clemson program we value the types of skills that students can carry away with them into the "real world": problem solving, data analysis, experimental design, and both oral and written communication skills. The traditional skills normally emphasized in a beginning laboratory course are taught during the lab, but in reality, and except for chemistry majors, these are not the skills most students will need later on. Thus lab techniques are introduced as a means to an end in the context of the experiment. It is also important to realize that these lab techniques should be taught and not merely "picked up" along the way. It is unfair to leave students without these skills, and they cannot and should not be expected to "construct" these skills in the same manner that they will construct the knowledge and less concrete skills that will be more useful to them later. The teaching of techniques in our course is accomplished with the help of an on-line resource, SuperChemLab, that was developed for this purpose. However, it is certainly possible to develop a successful program without the multimedia aids.

The lab course outlined in this manual has been in place at Clemson for the past ten years. The first two years were spent in evaluation of the course with half the students in regular labs and half in cooperative labs. During the past fifteen years all our science and engineering general chemistry labs have been taught this way. Realizing that many instructors might feel a little uncomfortable with such a whole-scale conversion to the new laboratory program, a useful beginning might be to include a project at the end of the first semester or in the second semester. However, if your experience is anything like ours, you will be impressed with the abilities and ingenuity of your students.

The projects outlined in this manual are quite demanding of the students. One common student comment was that these labs were more difficult and required more from them. If we required students to perform these projects individually, it is my feeling that very few of them would be successful. This is one reason why we used the educational technique known as cooperative learning as the support basis for these labs. Cooperative learning has its roots in K–12 education circles, and it is one of the most extensively researched educational techniques. In our course we use most of the tenets of cooperative learning:

1. Students must develop a positive interdependence; that is, "we all sink or swim together." This is particularly important when, as often happens during the first project, the students feel overwhelmed by what is being asked of them. It can be very comforting that "they are all in the same boat."

2. Social skills: The first lab period is spent developing a group camaraderie and making sure that students feel comfortable and can function in a group setting. We have also found it important to be up front with students in letting them know why we are teaching the laboratory this way. We tell them that we value the skills they can take with them and that we want them to have the opportunity to develop and practice their communication skills. There is a large section of the lab manual devoted to course policies and group functioning. It is our experience that for cooperative learning to work well it must be quite structured as outlined in the manual.

3. Individual accountability: In the long run it is the individual student's performance that must be graded. In these labs, and in the suggested grading scale, the majority of the grade comes from individual student's contributions. Some educators have suggested that group grades, particularly group lab reports, are a viable method of grading, but most practitioners of cooperative learning feel (as I do) that assigning a large portion of a student's grade from a group effort is a recipe for trouble. Complaints from good

students who are unhappy about other students getting credit for work that is not their own, and problems with students with poorer learning skills who might be tempted to hitchhike onto a better (or more hardworking) student's efforts, are to be expected if group grading is carried out.

An excellent resource for those who are interested in cooperative learning is *Active Learning: Cooperation in the College Classroom* by D. Johnson, R. Johnson, and K. Smith, published by Interaction Book Company, Edina, MN, 1998.

COOPERATIVE CHEMISTRY LABORATORIES

To the Student

Welcome to Cooperative Chemistry Laboratories! The format of this chemistry laboratory will probably be different from the one to which you are accustomed. Instead of performing one lab exercise per week on an individual basis, you will become a member of a group of students who will work on an assigned, open-ended problem for a number of weeks. In a typical semester, your group might complete three to six projects, rather than twelve pre-prepared experiments.

The idea behind this type of laboratory is to give you a more realistic experience in the laboratory. Real scientists do not have a recipe for each experiment that they perform. Rather, they devise their own experiments to test their hypotheses and gather as much data as they can. In the real world there is no one right answer, and so this should be in your laboratory experience as well. You will not be graded on how close you come to a "right answer," but rather on how you plan your experiments, evaluate your data, and present your report. The lab is **not designed** to coordinate topics with the lecture portion of the course. (Note that there is no evidence that laboratories that are designed to reinforce lecture topics actually result in deeper understanding.) However, you will need to apply knowledge gained from the lecture. In some cases, you may need to learn new material that will then be applicable to the lecture. While pursuing this lab course, we will be concerned with teaching you how to think your way through problems, a skill that will serve you well in any field.

A Word about the Things You Will Learn in This Course Besides Chemistry

A survey of major businesses and industrial firms concluded that the workplace basics to absorb in school are:[a]

1. Learning how to learn.
2. Listening and oral communication.
3. Competence in reading, writing, and computation.
4. Adaptability based on creative thinking and problem solving.
5. Personal management characterized by self-esteem, goal setting, motivation, and personal and career development.
6. Group effectiveness characterized by interpersonal skills and teamwork.
7. Organizational effectiveness and leadership.

[1]American Society for Training and Development and the U.S. Department of Labor (1988).

Unfortunately, many of these things traditionally are not taught in colleges and universities. However, if you as students are to reach the workplace well equipped to cope with the real world, you must have the opportunity to practice some of these skills. Learning these skills will be one of the major goals of this laboratory in addition to learning about how chemicals react and some laboratory skills. You will be presented with problems that you must solve in the context of the general chemistry lab. When you begin this process, it is very likely you will experience some frustration, and perhaps get discouraged, but remember: **BEING TEMPORARILY PERPLEXED IS A NATURAL STATE OF PROBLEM SOLVING.**

The feeling you get when faced with a new problem and do not know what to do is quite normal. The key here is that the sensation is temporary and that as you get into the project you will begin to feel more confident. By the end of the course you will have learned a set of problem-solving skills that you can carry with you into the rest of your life. At the beginning of each of your projects, the time spent in which you may seem to be floundering is in fact not time wasted. Rather, research has shown that this time is essential to successful problem solving and sometimes is given the term "creative floundering." After all, if you knew immediately how to solve a problem, then it would not really be a problem in the true sense, merely an exercise.

This lab course will present you with some problems that you may not be able to solve alone. That statement brings us to the other rather unique aspect of this course: you will be assigned to a group of students, sometimes known as a cooperative learning group.

COOPERATIVE LEARNING

Much of the work that you will do in this laboratory course will be done as teamwork; therefore, it is important that you understand some of the goals and benefits of group learning (or cooperative learning, as it is sometimes called). For some of you this may be the first time that you have been exposed to group learning, so we will begin the laboratory session with a group learning exercise. This should enable you to get some feel for the kind of interactions that occur.

Cooperative learning is the name given to a style of learning that has become quite prominent during the past few years. There is considerable evidence that the outcomes from cooperative learning are very positive for the student. Research has shown that students learn better and with more understanding, that they develop interpersonal skills, and that they enjoy the course more when it is conducted in this style.

Cooperative learning is not simply placing students in groups and having them work side by side on a problem. Nor is it having the best student do the work while the others watch. The elements of cooperative learning are the following:

1. **Positive interdependence.** That is, you all "sink or swim together." Everyone in the group should pull his or her own weight. Under this system everyone has something

to contribute, and everyone will be able to contribute something. The more you put into the group's efforts, the more you will be rewarded at the end, not only in terms of the grade you receive but also the skills you learn and the satisfaction you get from your accomplishments.

2. **Individual accountability.** Although the projects will be carried out by the group, the evaluation of your performance will be individual. You will be graded individually on your lab reports, lab technique, and peer evaluation. Only a minor component of the grade is a group grade. This is for two reasons: (a) to discourage "hitchhikers" (that is, students who do not contribute to the group), and (b) to ensure fairness so that students will get the grade they earned rather than the grade someone else earned for them, be it higher or lower.

3. **Social skills.** Groups cannot function effectively if students do not have or learn leadership, decision making, trust building, communication, and conflict management skills. During the semester you will be assigned different roles with different responsibilities. These roles will be rotated after each project so that you will each have a chance to take up most of the roles. At the beginning of the semester you will perform several exercises designed to develop some of these skills. Others will develop as you have a chance to use them.

CONFLICT MANAGEMENT

When people work in groups, it is important that they be able to communicate with one another without conflict. It is inevitable that sometime during the course of the semester one of your group members will say or do something that will annoy you or that you disagree with. So it is important to bear in mind the following:

- **Be critical of ideas not of people.** When you are discussing your plan of action for the project, it is almost certain that conflict will arise. If someone has an idea that you think is stupid, it is OK to criticize the idea (it may not be a good idea to call it stupid), but if you criticize the person, it is almost certain to cause hard feelings and affect the functioning of the group.
- **Try to avoid win-lose situations.** The goal of this lab course is to develop problem-solving skills, not to engage in conflicts in which one person's ideas dominate.

During this lab course you will also learn, or be given the opportunity to practice, your **oral and written communication skills**, since you will be required to produce both written and oral project reports.

We would like you to assign one of the following roles to each member of the group. Note that it will be possible to change your role in the group later in the semester, but you will keep your role for the first project:

Member A—Team Leader. Responsible for overall supervision of the team and for ensuring that all members contribute equally.

Member B—Communicator. Primarily responsible for communicating with the instructor.

Member C—Record Keeper. Keeps records of all materials discussed and data collected; is responsible for informing absent team members of progress made.

Member D—Team Counselor. Responsible for making sure all team members agree on the proposed course of action; also responsible for completing all group submissions in a way that reflects the thinking of all team members.

Remember that although much of your project will be designed and carried out by the group members, your grade will be determined mainly by your individual efforts. Therefore, it is important that you do not let the other members of the group do your work and that you participate to the fullest extent. As you plan your experiments, it is important that everyone is occupied most of the time. The idea is not that one of you will do the experiment while the others watch, but that each of you will contribute to the overall goal. This may mean that some of you will do duplicate experiments or that each member of the group might perform a given experiment under a different set of conditions. As you devise your preliminary plan, it is important to remember to give tasks to everyone on your team.

NATURE OF THE COURSE

As already stated, this course is probably quite different from those that you are used to. Although you may feel somewhat unsure of what to do initially, remember that your group is there to help. You are all in the same boat and can look to each other for help. We hope that you will find these labs fun, interesting, and intriguing. If past experience is anything to go by, this will be true. Remember, there is no one right answer for any of the projects you will be assigned. There is no set amount of material to be covered, and there is no point at which you should stop and say you have done "enough." Each group in your section will be doing different but related projects to yours, and each will have different answers. It is up to your group to define the problems you are given and how far you will go to obtain the solutions.

BRIEF OUTLINE OF THE COURSE

Typically during the first lab period, you will be assigned to a group and spend some time getting to know the members of your group and acquainting yourselves with the safety rules and the laboratory equipment. You will then be given a number of tasks to accomplish for the purpose of familiarizing you with a few key lab procedures. The first project will be introduced, and you will spend the rest of this laboratory period preparing a plan of action and practicing laboratory skills.

At the beginning of the second lab period you will begin experimental work on the first project, and you may need to change your plans depending on how your experimental work proceeds. Most projects will last for two to four weeks. So at the end of each lab period, you should spend some time in discussion of that week's work and what you will do next week. Typically your grade for the lab will result from a number of factors including a written lab report, an oral report or poster presentation, your laboratory notebook, and your performance in the lab as evaluated by both your peers and your instructor. The specifics of how you will be evaluated may change, but what follows are guidelines for reporting your data in various ways.

RESOURCES

The resources available to you in this course are numerous.

1. **Your laboratory instructor.** The role of your lab instructor may be somewhat different from what you are expecting; in some lab courses the instructor begins by telling students what to do and how to do it. Since the philosophy of this course is that you will learn more if you design your own projects and experiments, the instructor will not have all the answers to your questions. After the instructor has explained the nature of the course to you, you will be required to devise a plan to get started on each project. Your instructor is not in lab to tell you what to do and will not answer questions of that nature. However, he or she will be available to guide you in the right direction. Be assured the instructor will not let you head off in too many unproductive directions. The role of your instructor is more of a "guide on the side" and safety officer. You and your group must take the responsibility to plot your course, design your own experiments, analyze your data, and then make sense of all the information that you have gathered. From time to time your instructor will have suggestions and helpful hints for you, but again do not expect the instructor to tell you explicitly what to do.

2. **Your group members.** When you begin the lab, you might feel a little overwhelmed, but remember there are people in your group who have different ideas and experiences than those you have had. They may have insights and ideas about the project that you may not, and vice versa. Don't forget that your group members can be a valuable resource. Be sure you listen when people have something to say.

3. **Textbooks.** One of the purposes of this lab is to allow you to use the knowledge that you have been learning in lecture. We want this lab to be as much like real life as possible, and that means that you may use as many resources as you can to support your research effort. Probably the first place you will want to look is your textbook. It has a great deal of information in it, arranged in a way in which you are probably familiar, so this should help you know where to look. In addition, most textbooks have tables of constants and thermochemical data. It is a good idea to bring your textbook to lab every week.

4. **Reference materials available in the laboratory.** There are probably a number of catalogs and general reference works such as the *CRC Handbook of Chemistry and Physics* available in the lab. These will give you an additional source for information

about specific compounds' physical constants, thermochemical data, or the makeup of specific solutions.

5. **Libraries, computers, and on-line searching.** It is highly unlikely that your textbook will contain everything you need to know about a particular topic or compound.

SAFETY RULES

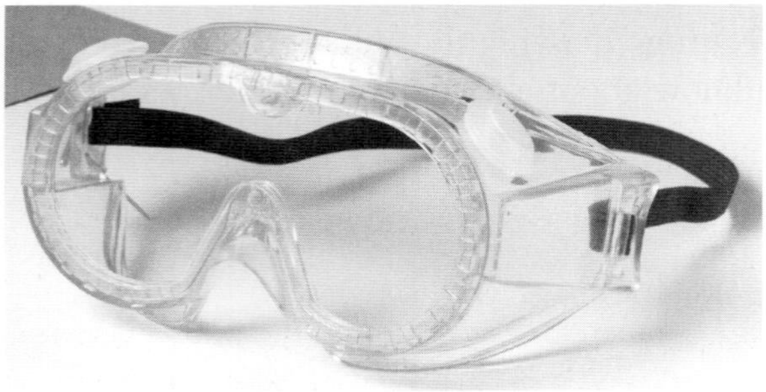

© The McGraw-Hill Companies, Inc./Ken Cavanagh Photographer.

1. **Wear your goggles.** Wear approved eye protection at all times while in the laboratory and in any area where chemicals are stored or handled. Eye protection should protect against impact and chemical splashes. Goggles that conform to the **ANSI Z-87.1, 1989** standard are required. Prescription glasses and sunglasses are not acceptable forms of eye protection. If you purchase goggles, make sure that they have the **ANSI Z-87.1, 1989** imprint on them and that they are designated as chemical splash goggles.

   Under no circumstances should you wear contact lenses in the laboratory, even under goggles. Chemical vapors may dissolve in the liquids covering the eye and concentrate behind the lenses. "Soft" contact lenses are especially dangerous because chemicals can dissolve in the lenses themselves and be released over a period of several hours.

   The laboratory has an eyewash station for your use. In the event that a chemical splashes near your eyes, use the station before the material runs behind your safety glasses and into your eyes. **You should irrigate your eyes for at least five minutes and notify your laboratory instructor.**

2. **Don't ingest anything.** Eating, drinking, and smoking are prohibited in the laboratory at all times.

3. **Dress like a chemist.** Wear clothing in the laboratory that will provide maximum body coverage. Shorts, mini-skirts, etc., are completely inappropriate. Your clothing should come down to your ankles. Wearing open-toed shoes and sandals is also not allowed in the lab.

   It is advisable to wear old clothes in the laboratory in case of spills. You might also want to keep a pair of sweat pants in your drawer. Long hair should be securely tied back to avoid the risk of setting it on fire or contaminating it with chemicals.

   If large amounts of chemicals are spilled on your body, remove the contaminated clothing and use the safety shower located in your laboratory. Stand under the shower for at least five minutes, and make sure that your laboratory instructor is aware of the problem.

4. **Don't touch any chemicals.** Never taste or touch any chemical. Many chemicals are absorbed through the skin. Wash off all chemicals with large quantities of running water. If directed to smell a vapor, gently waft the vapors toward your nose. Do not smell the source of the vapor directly.

5. **Use the hood.** Perform any reactions involving toxic, irritating, or otherwise dangerous chemicals or unpleasant odors in the hood. Each student should have access to a hood.

6. **Know what to do if there is an accident.** In case of fire or an accident, call the instructor at once. Note the location of the fire extinguishers, safety showers, and safety blankets as soon as you enter the laboratory so that you may use them if needed.

   a. Wet towels can be used to smother small fires.

   b. If a person's clothing catches fire, pull out the fire blanket and roll the person up in it to smother the fire.

   c. In case of a chemical spill on your body or clothes, stand under the safety shower and flood the affected area with water. Remove contaminated clothing to prevent further reaction with the skin.

NOTIFY YOUR INSTRUCTOR IN CASE OF ANY ACCIDENT, NO MATTER HOW SLIGHT.

7. **Work safely.** Do not point a test tube at anyone while you are heating it; it may erupt.

8. **Beware of glass; it is breakable.** When inserting glass tubing or thermometers into stoppers, lubricate both the tubing and the hole in the stopper with water. Wrap the tubing in a towel; grasp it as close to the end being inserted as possible; and push gently, using a twisting motion.

9. **Assume you are the only safe worker in the lab.** Work defensively. Do not assume that everyone else is as safe a worker as you are. Wear your goggles even when you are not working. Be alert for others' mistakes.

10. **Do not wear headphones in the lab.** Anything that interferes with your ability to hear what is going on in the lab is a potential hazard.

BASIC LABORATORY ETIQUETTE

The equipment you use in this laboratory is used by many other students. **Please leave the equipment and all work spaces as you would wish to find them.**

1. If you find a piece of equipment that is not in good working order, notify your teaching assistant (TA) immediately so that a work order for its repair can be filled out.
2. After you have finished work in the lab, clean off your work area so that it is in good shape for the next person in the lab. It is a good idea to bring a roll of paper towels and some liquid soap to keep in your lab drawer.
3. When obtaining reagents, do not take the reagent dispenser to your desk.
4. If you take more reagent than you need, **do not** put the excess back into the bottle. It may be contaminated at this stage. Treat the extra as waste and dispose of it accordingly.
5. You will be given instructions on how to dispose of waste chemicals on the basis of their toxicity. Do not put anything down the sink unless you are explicitly told to dispose of it this way. Most waste will be placed in labeled containers.
6. When weighing any material on the balances, do not weigh directly onto the balance pan. Weigh your material on a piece of weighing paper or into a beaker or other container. If you should spill anything onto the balance, notify your TA or the stockroom manager immediately. The balances are sensitive instruments and should be treated with great care.
7. Do not put stoppers from reagent bottles down onto the lab bench. They may become contaminated. Hold the stopper in your other hand while you get the material out of the bottle. Replace stoppers on the bottles they came from to prevent cross-contamination.

Above all, **IF YOU MAKE A MESS, CLEAN IT UP OR TELL SOMEONE ABOUT IT. DO NOT LEAVE IT FOR SOMEONE ELSE TO FIND.**

NFPA HAZARD CODES

NFPA stands for the National Fire Protection Association. The association originally developed the following set of hazard rankings for their own purposes, but the rankings have proven to be very useful in the chemical industry.

Flammability

0 Material will not burn.
1 Material must be preheated to burn.
2 Material must be preheated or exposed to relatively high ambient temperature to ignite.
3 Material can be ignited under almost all ambient temperatures.
4 Material will rapidly or completely vaporize readily at standard temperature and pressure (STP) or will burn readily when dispersed in air.

Reactivity

0 Material is normally stable and is not reactive with water.
1 Material is normally stable but may become unstable at elevated temperatures or may react with water.
2 Material is normally unstable and readily undergoes violent chemical change but does not detonate. May react violently with water or may form potentially explosive mixtures with water.
3 Material is capable of detonation or explosive reaction but requires a strong initiating source or may react explosively with water.
4 Material is readily capable of detonation or explosive decomposition or reaction at standard temperature and pressure (room pressure, 0°C).

Toxicity

0 Exposure to materials under fire conditions offers no special health hazard.
1 Exposure to material can cause short-term irritation or minor residual injury if untreated.
2 Intense or continued exposure to material can cause temporary incapacitation or possible residual injury unless prompt medical attention is given.
3 Short exposure to material can cause serious temporary or residual injury even if prompt medical attention is given.
4 Very short exposure to material can cause death or major residual injury even if prompt medical attention is given.

Other Hazard Codes

1 OXY—an oxidizing material
2 Corrosive—a strong acid or base
3 Radioactive material
4 Hazardous material when in contact with water

WASTE DISPOSAL

After you have finished your experiments, you will probably have some waste chemicals and solutions to dispose of. **Do not put them down the sink unless specifically told to do so by your instructor.** It is your responsibility to investigate the appropriate waste disposal methods by consulting the Material Safety Data Sheet (**MSDS**). There will be receptacles for all types of waste in the lab including **reactive, corrosive, flammable,** and **poison.**

MATERIAL SAFETY DATA SHEETS

An MSDS is an information sheet that contains all the safety and waste disposal information on a particular compound. Sample MSDSs are available in the lab. You will be required to check these sheets before dealing with any compound in the lab. If the MSDS is not available, you should be able to access the MSDS file from the Internet. Note that even a rather innocuous compound like sodium chloride has some toxicity and hazard information data.

Typically the MSDS will list the name, formula, registry numbers, and physical and chemical properties of the compound. Safety precautions to be taken when handling the material and what to do if there is a spill or accidental ingestion are also included.

The LD50 (the lethal dose for 50% of the population) gives an indication of the toxicity of the compound. For example, the LD50 for NaCl is 3000 mg/kg (oral dose in a rat), while the LD50 for NaCN is 6.8 mg/kg.

RECORDING AND REPORTING RESULTS

The Laboratory Notebook

In contrast to other laboratories you may have experienced, you will be required to keep a laboratory notebook for the duration of this course. The notebook should be bound, with the pages numbered consecutively for easy reference. The preferred notebook is the carbonless copy type—in which each page is reproduced twice so that both you and your laboratory instructor can keep a copy of your lab notes.

The notebook is a day-to-day record of your activities in the lab. It is the place where you will describe experiments **as you do them** and note observations **as you make them.** It is where you will record and analyze your data. Your notebook will be an invaluable tool throughout the semester because you must have an accurate record of what you did and what you observed in the laboratory when the time comes to write your laboratory reports.

All data, results, weights, etc., should be **recorded directly into your notebook in ink**. Loose pieces of paper might easily get lost. Remember that it is to your benefit to keep a detailed notebook because doing so will make it easier for you to reconstruct the experiment accurately in a report later. Any mistakes should be crossed out so that they can still be read. They should not be erased or removed with correction fluid.

The notebook should **not** be a neatly copied reiteration of the laboratory procedure. Rather, you should write what you did and observed. Neatness, spelling, punctuation, and grammar are not essential in this notebook, but it should be possible for someone else to repeat your work by reading your account; i.e., it should be legible and intelligible.

Your lab instructor will read your notebook periodically in order to see whether you are keeping up with your work, whether you understand what you are doing, and whether you are recording everything you need. Since the lab notebook should be used for recording everything that you do in the laboratory, there is not one particular format that can be used for all situations, and it is important to be flexible. However, there are certain things you should do to make the lab notebook as useful as possible.

- Leave a few pages blank at the beginning of the notebook. As you begin new experiments and projects, you can use these pages to prepare a table of contents that can be updated as you go.
- Make sure all the pages are numbered.
- Make sure that each page is signed and dated; although this may not be so important in an introductory lab, it is certainly good practice and is required in many research labs.

A sample from a lab notebook is given on the following page.

January 1, 2012
Experiment: Esterification of 3-nitrobenzoic acid
Week 2:
Purpose: To make an ester using 3-nitrobenzoic acid as the starting material
Materials used
3-nitrobenzoic acid 1.0 g
Methanol 25 mL
Conc H_2SO_4

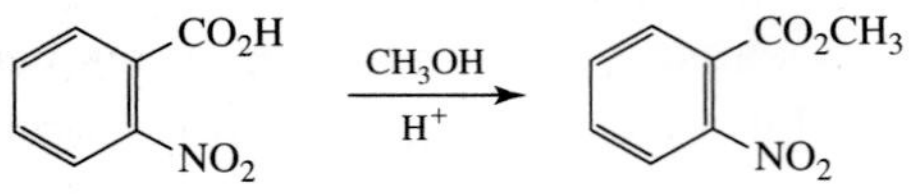

Reaction

1 g acid dissolved in MeOH, 2 drops H_2SO_4 added—the reaction mixture warmed up and darkened a bit.
I heated the reaction for about 20 minutes, and I took a sample for TLC—
TLC solvents tried: methanol, toluene, ethyl acetate. Best separation of reactants and products with toluene/ethyl acetate 9:1. After 30 minutes—no more starting materials.

TLC plate

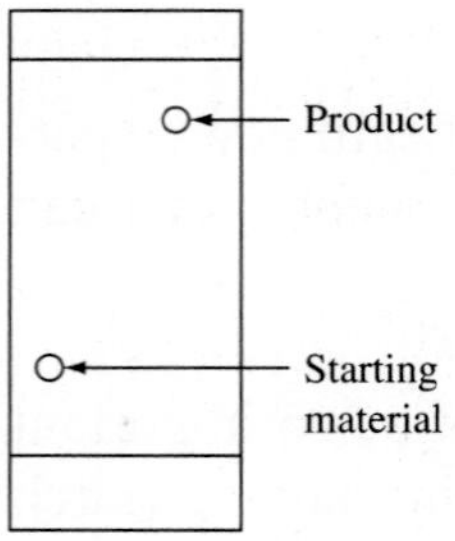

Reaction was poured in 0.1 M NaOH soln

- cream colored xtals pptated.
- Filtered dried
- Mass watch glass 32.08 g
- Mass watch glass + xtals = 32.58 g
- Mass xtals = 0.50 g

January 1 2012, Jane Smith

Writing Lab Reports

The following is an outline of the fundamentals of writing a lab report. Please read it carefully. There is also a sample laboratory report with comments for you to read. The grading criteria for your lab reports will refer to the following outline.

Scientists are investigators who "try out" ideas. They conduct experiments in order to test or prove ideas, and they share the results of their experiments in papers and written reports. Reports allow others to learn the results of scientific investigations. Like other scientists, you will be conducting experiments, stating hypotheses, observing processes, recording data, and formulating conclusions. For some of these experiments, you will be asked to write a **lab report** describing the experiment, summarizing your observations, and explaining your conclusions or judgments about the meaning of what you observed.

Audience

Professional scientists write papers and reports for their own use, for their colleagues, and for the public. When they write, they choose the language, style, and format best suited for communication with their readers. In preparing your lab reports, think of the ***reader*** as an educated person who is interested in learning about your experiment but who knows less about this subject than yourself. You take the role of expert; you are in charge of efficient, accurate communication of facts and ideas.

Evaluation

Lab instructors, of course, will actually be reading your papers, commenting on them, and grading them. Your instructor assumes the role of the reader described above. He or she judges how well you communicate necessary background information as well as the processes and results of the experiment itself. Your instructor also observes how carefully you follow the format for lab reports described in this manual. The grade sheets included in this booklet indicate some of the things looked for by lab instructors. Read them carefully, and ask your lab instructor about any remarks or corrections that puzzle you. Make sure you understand what the criteria on the grade sheet mean and what parts of the lab report they apply to.

General Format

Lab reports and papers must be typed or word processed. They should include tables and illustrations where necessary. Lab reports contain the following sections: ***title page, abstract, introduction, results, discussion, conclusion, experimental,*** and ***references.***

Title Page: The first page of the report is the title page. Include the title of the report, your name, the course number, your section number, the instructor's name, and the date the report is due. In addition, you should include a statement that indicates that this paper is your own work. Put the information in the order just presented. *Note:* The title of the paper can simply be the name of the experiment. For example,

The Kinetics of Solvolysis of *t*-Butyl Chloride
Jane Doe
Chemistry 101 Laboratory, Section 255
Instructor: Linus Pauling
January 2, 2012

or for a more complex project it can be an overall description. For example,

An Investigation of the Structure and Properties of an Unknown Compound

You should also include a statement indicating that you have not committed an academic integrity violation. For example,

My signature indicates that this document represents my own work. Excluding shared data, the information, thoughts, and ideas are my own, except as indicated in the references.

Your laboratory instructor will clearly explain which areas of the report (if any) may be shared with group members.

Abstract: The abstract gives a condensed version of the whole report. It should give readers an idea of what was done and why, in a few sentences. Typically the abstract should be between 100 and 200 words. In practice, the abstract will be the last thing that you write. When you have everything that you have done organized and explained, it will be easier to write a condensed version.

Introduction: The text of the report begins with an introduction. In this section of the paper, identify the experiment(s). In general terms, tell the reader what you intend to do and why you intend to do it. Include all phases of the experiment(s). Emphasize any unusual or critical conditions.

Point out exactly what it is you are trying to do. Make sure the reader understands the purpose of the experiment(s). What you are attempting to test or prove or investigate should be clear to the reader. If necessary, include general information that explains the importance of the experiment(s) and why you are doing this project. You can use the guiding questions after each experiment to help you organize your thoughts.

Even though you are acting as the "expert" for the general reader, you must cite references both to support statements you make about the scientific basis for your investigation and to define sources for specific pieces of data crucial to the experiment. Citations give you credibility: they tell the reader that you have prepared properly for the experiment by providing yourself with a basic background and by "checking out" the accepted authorities. These citations should be numbered consecutively in the text and listed in the **References** section (see **References** at the end of this section).

Typical sources for your citations include your textbook, other lab manuals you may have consulted, tables of data such as the *CRC Handbook,* chemical catalogs, and Internet resources.

Results: In this section, you summarize the outcome of your experiments for your reader. This section will consist primarily of **data** (facts and figures) that you gathered in the course of the experiment. Read the section of the lab manual on reporting numerical results. Data must be presented in such a way that they are easy to read. You must organize or assemble and label the data for the reader. Numerical data or lists of numbers are usually presented in tables. Relationships between sets of data or factors in the experiment are often shown in graphic form. Graphs, drawings, and sketches are called **figures.** Although tables and figures are labeled on the page with descriptive titles, they are identified in the written body of your report by number rather than name. When you discuss tables and figures in the text of this section, you mention Table 1, Table 2, Figure 1, and so on.

Place tables and graphs at appropriate points in the body of your report. This makes it easy for the reader to use and understand the graphs, charts, tables, etc. If you put the data presentation in a separate appendix, the reader will have great difficulty in understanding the results of your experiment. It is permissible to have one large section called Results and Discussion, in which the results are presented in tables, graphs, and charts and then discussed in a block of text immediately following.

Discussion: The discussion section of your report is the most important one for you and your reader. In this section of the report, you interpret the results of your experiment for the reader. You explain what the results mean, how you know what they mean, and why you can make these claims—that is, you will discuss how the experimental evidence supports your discussion. You will also mention any weaknesses or problems in the plan of the experiment or methods you used. You demonstrate not only how successful your experiment was but also how well you understood the experiment. The discussion section can be difficult to write, but you will learn more about your experiment and yourself as an investigator as you write it.

Before you begin writing this section, complete the ***Introduction, Results,*** and the ***Experimental*** sections. Put these sections on the desk together with your lab notebook so that you will be able to look at all these sources of information as you write the ***Discussion*** section. Then prepare to write a rough draft of this section. You can make an outline for yourself by taking the following steps.

1. Write out your ideas and goals again. What were your questions? Look over the tables, figures, and general information you compiled for the ***Results*** section. What did your experiments show?

2. What did you do? Write down the specific data that led you to any conclusions you have come to.
3. What did you see? What observations did you make?
4. What can you claim from the results of your experiments? How do you know this? Can you defend your claims? Does the experimental evidence back up your claims? Write down what you know about the principles of chemistry involved in your experiment. How do the results of your investigation fit with the chemical concepts that you have learned? Identify the sources of your information at this point.
5. List any weaknesses or problems you discovered in your experimental design or procedures. Tell the reader how these problems may have affected the results of your experiment.
6. Review the experiments again. List and discuss any difficulties that arose during the course of the experiment. Be sure to point out to the reader any way in which these problems could have affected the outcome of the experiment.

Using this list as a guide, prepare a comprehensive discussion or explanation of the results of your experiments.

Conclusion: Your overall conclusions about the project have probably already been mentioned several times in the course of your report. You may have predicted some of the outcomes of your experiments in the ***Introduction*** and discussed them again as an empirical conclusion (meaning that it was derived from your experiments and observations) in the ***Results*** section. In the ***Conclusion*** section, a brief, single paragraph may be enough to clearly state the outcome of your investigation. The ***Conclusion*** section tells the reader what the results of the experiment mean. In a sense, it is a summary of the ***Results*** and ***Discussion*** sections combined. Make sure your stated conclusions clearly match the actual outcome of your experiment(s).

Experimental: In this part of the report, give the reader a step-by-step account of the actual experiment. Here, you will do more than simply provide greater detail about the experiments than you did in the ***Introduction.*** You need to describe your procedures in such a way that others could read your lab report, follow your lab procedure, and successfully duplicate your experiment. Scientific experiments are not considered valid unless they can be repeated by other experimenters working in other laboratories. In one sense, science has no secrets: scientific theories become established only when the experiments that led to them can be repeated or verified by others besides the original investigators. You make this verification possible when you write a complete, accurate description of your experiment. Scientists also use lab reports as a means of learning and sharing techniques. Other investigators may not choose to duplicate your experiment, but they may choose to use your procedure in some similar investigation. The ***Experimental*** section of your lab report should be usable as a set of directions for other scientists.

As you are writing, pay close attention to your style. The style used in scientific reports is somewhat different from that required for an English paper. It is important that you understand and get used to using different writing styles appropriately. Generally, you should write the experimental section in the past tense (you are reporting what you did, not what you are doing now). Write "the solution was filtered," rather than "filter the solution."

When reporting quantities, you should be as specific as possible. The usual style is to write the name of the substance you are using followed by the amount in parentheses. For example: sodium hydroxide (4.0 g, 0.10 mol), not 4.0 g sodium hydroxide.

The ***Experimental*** section can be placed after the ***Introduction*** if you wish; however, you will avoid the trap of discussing results if you place it last as it is found in most chemistry journals.

References: The final section of your report tells the reader where to find any of the sources of information you used in your report. In the body of your report (particularly in the ***Discussion*** section), you may have mentioned other texts. Each of these references should be numbered consecutively within the text as superscripts. At the end of your report, after the ***Conclusion***, include a complete reference list, in numerical order, of the sources used. The reader can use this list to follow up or check out any source you mentioned or to do additional reading.

The format of the references should follow the American Chemical Society Guidelines, which can be found in the *ACS Style Guide.* For example,

A similar experiment has been reported by Haight[1] and expanded by Vogel.[2]

[1]Haight, G. P. *J. Chem. Educ.,* **1965**, *42*, 468.
[2]Vogel, A. I. *A Textbook of Qualitative Inorganic Analysis,* Longman: New York, **1979**, p. 358.

In general, the format for journals is as follows:

author (last name, initials), *title (in italics)*, **year (bold)**, *volume number (italics)*, starting page number

The format for books is as follows:

author (last name, initials), *title (in italics)*, publisher: city, year, page number

For multiple authors, separate the authors by a semicolon. For example,

Corey, E. J.; Weinshenker, N. M.; Schaaf, T. K.; Huber, W. *J. Am. Chem. Soc.*, **1969**, *91*, 5675.

For further information see Reference Style Guidelines at http://pubs.acs.org/books/references.shtml.

If you find a reputable resource on the Internet, then you may also cite that as a reference. Be careful in the sites that you visit; try to stay with sites hosted by professional societies, the government, or other reputable institutions.

When citing an Internet site, you should give the URL, the host institution, and the date of your visit to the site (since many Internet sites are transient). For example,

SuperChemLab. http://chemed.eng.clemson.edu/SCL/index.html, Clemson University (accessed January 1, 2008).

A sample lab report follows this section.

Preliminary Lab Report Guidelines

Most teachers of writing stress the importance of rewriting and editing. In order to give you experience in this area you will probably be asked to write a preliminary lab report. This exercise will give you the chance to write a fairly brief report and get feedback **before** you write the major report on your project. In this way you get to make any mistakes on your preliminary report, rather than on the full report.

If you are asked to write a preliminary report, read the report guidelines first and use the same general format. Usually you will not have all your results, and you may not have any conclusions at this stage. However, you can include the other sections, with as much detail as possible.

The Science Writing Heuristic

In some instances, your laboratory instructor may ask you to write a less formal report. If this is the case, the science writing heuristic is designed to help you as you organize your thoughts and develop answers to the most important questions you should be asking as you write the report.

The science writing heuristic consists of a number of organizing questions.

1. **Beginning ideas—What are my questions?** Use the guiding questions after each project to help you organize what you were trying to do in the experiment.
2. **Tests—What did I do?** Use your lab notebook to help you record and remember what you did.
3. **Observations—What did I see?** Make sure that you record all observations in your lab notebook as you go along.
4. **Claims—What can I claim?** Make sure you make claims that you can defend with experimental evidence.
5. **Evidence—How do I know? Why am I making these claims?** Explain exactly why it is that you can make the claims you are making, and give the evidence for why you are making these claims.
6. **Reading—How do my ideas compare with other ideas?** How does what you have written compare with relevant material that you may have learned in the lecture or you have found in your background reading?
7. **Reflection—How have my ideas changed?** Write a paragraph about what you have learned from this experiment. How do your ideas about this subject compare now with ideas you had before you performed the experiment(s)?
8. **Writing—What is the best explanation for what I have learned?**[b]

[b]Takett with permission from B. Hand and T. Greenbowe, "Introduction to Science Writing Heuristic," chap. 12 in *Chemists' Guide to Effective Teaching,* ed. N. Pienta, M. Cooper, and T. Greenbowe, Prentice Hall: New Jersey, 2005.

An Investigation into the Structure, Properties, and Synthesis of 3-Nitrobenzoic Acid

Jane Smith

Chemistry 101 Laboratory, Section 43
Instructor: Marie Curie
January 2, 2012

Abstract

This paper describes the identification and investigation of the properties of an unknown compound. The compound was identified as 3-nitrobenzoic acid by a combination of tests. The compound was an organic compound with a melting point of 131–141°C, which was found to be most soluble in polar organic solvents. Since the compound was more soluble in sodium hydroxide solution than water, it was deduced that the compound was an acid. Quantitative solubility estimates of the compound and a molecular weight found by titration also supported this hypothesis.

The compound was esterified to produce the methyl ester that exhibited the expected loss of acidity and decreased solubility in base.

Introduction

Note that the introduction is quite short. Remember brevity can be the essence of good communication.

The goal of this laboratory project was to investigate the structure and properties of an unknown compound that had been found in an unmarked bottle in the high school laboratory. The team of investigators was called in when a new high school chemistry teacher noticed the bottle sitting in the back on a shelf. It was important that the compound be identified and its physical and chemical properties identified so that it could be disposed of properly. In addition it may be possible to use this compound for other purposes. The synthesis of the unknown compound was also a goal of this project.

Note that each table is given a number and a descriptive title.

Results

TABLE 1 Physical Property Tests

Appearance of compound:	Cream-colored needles
Smell	None
Melting point	139–141°C

TABLE 2 Qualitative Solubility

Solvent	Solubility	Color of Solution
Toluene	Slightly?	Pale yellow
Acetone	Very soluble	Pale yellow
Methanol	Very soluble	Pale yellow
Water	Slightly?	Pale yellow
Dilute HCl	Insoluble	Pale yellow
Dilute NaOH	Very soluble	Deep yellow

TABLE 3 Chemical Tests

Test for aldehyde or ketone	Negative
Test for phenol	Negative
Test for alcohol	Negative
Test for chloride	Negative
Test for sulfate	Negative
Test for carbonate	Negative

TABLE 4 Titration Results

Run	Wt of Unknown	Vol 0.100 M NaOH	m wt
1	0.242 g	14.49 mL	267
2	0.251 g	15.12 mL	266
3	0.248 g	15.00 mL	265

Sample Calculation

$$14.49 \text{ mL} \times \frac{0.100 \text{ M NaOH}}{0.100 \text{ mL NaOH}} \times \frac{1 \text{ mol acid}}{1 \text{ mol NaOH}} = 1.449 \times 10^{-3} \text{ mol acid}$$

$$\text{Molecular weight of acid} = \frac{0.242 \text{ g acid}}{1.449 \times 10^{-3} \text{ mol acid}}$$

$$= 167 \text{ g/mol}$$

$$\text{Average m wt of acid} = 166 \text{ g/mol}$$

$$\text{Actual m wt of 3-nitrobenzoic acid} = 167.12 \text{ g/mol}$$

$$\text{percent error} = \frac{167 - 166}{167} \times 100$$

$$= 0.6\% \text{ error}$$

TABLE 5 Absorbance-Concentration Data

Concentration	Absorbance
8.00E − 04	1.78
6.00E − 04	1.335
4.00E − 04	0.89
2.00E − 04	0.445
1.00E − 04	0.222

The graph was pasted in from another document.

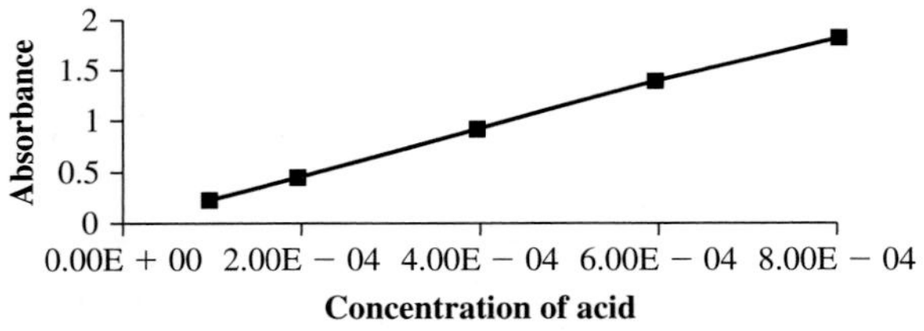

Figure 1: Calibration Curve

The slope of the line was found to be 2.25×10^{-3} by using the LINEST function in Excel. If time had permitted, the slope of the calibration curve would have been used to calculate the concentration of the acid in the samples removed from the hydrolysis reaction.

Discussion

The unknown compound was composed of yellow crystals that had the form of long needles when examined under a magnifying glass. The compound was odorless as indicated in **Table 1**.

Solubility tests indicated that the compound was very soluble in both acetone and methanol and slightly soluble in water and toluene as indicated in **Table 2**.

This information indicated that the compound was quite polar in nature since it was not very soluble in toluene, which is a nonpolar solvent. The compound was not soluble in water, which would rule out many of the soluble ionic salts. In fact all ionic compounds could be ruled out since the compound dissolved in acetone and methanol. This led to the conclusion that the compound was a polar molecular compound.

The compound was insoluble in dilute acid (HCl); however, it was soluble in dilute base (NaOH). Solubility in base implies that the compound must have acidic properties. In order to investigate this property further the compound was treated with a 1 M solution of Na_2CO_3, which is a weaker base than NaOH, in which it was also soluble. This means that the compound is quite a strong acid, since it is soluble in a relatively weak base.

A determination of the quantitative solubility of the compound showed that the solubility was 22 g/L in 1.0 M NaOH. The pH of a slurry of the compound was tested and found to be about 4. At first the compound was thought to be insoluble in water, but some must have gone into solution, otherwise the pH of the solution would have been 7. (The pH of pure water was measured with the same pH meter to make sure the meter was functioning properly.)

The tests of physical properties gave some insight into the nature of the compound, but further chemical tests were needed to identify the compound. As indicated in **Table 3,** tests for common ions such as chloride, sulfate, and nitrate were negative, as were tests for organic functional groups such as aldehydes and ketones, alcohols, and phenols. These tests ruled out a number of compounds as possibilities but did not give any further information about the compound.

The infrared spectrum of the compound did provide more information about the compound; a broad band from around 2500 to 3200 cm^{-1} indicated a carboxylic acid group. This information reinforced the finding that the compound was acidic. In addition, a strong band at 1620 cm^{-1} indicated the carbonyl group of the carboxylic acid. Two bands around 1530 and 1320 cm^{-1} seemed to indicate the presence of a nitro group. Since two of the possible compounds for the unknown were 3- or 4-nitrobenzoic acid, it was possible to narrow down the viable candidates for the unknown to these two compounds. The melting point of the compound was 139–141°C. The literature value[1] for 3-nitrobenzoic acid was 140–142°C, while that of the 4-nitro isomer was 239–241°C. From this information it was concluded that the compound was 3-nitrobenzoic acid.

These tests and spectroscopic data indicated that the compound had been identified qualitatively, but a quantitative proof of identity was needed to ensure that a correct identification had been made. Since the compound is an acid, a titration should give the information needed.

The compound was not very soluble in water, but if 0.25 g acid was heated in about 200 mL of water it would go into solution. (At first the titration was attempted with methanol as a solvent since the acid is much more soluble in this solvent; however, as indicated in Tro[2] pH titrations are only valid in aqueous solutions.) Since all the pH meters were in use, an indicator was used to identify the endpoint of the titration. An indicator that would change in the pH range of around 9–10 was needed, so phenolphthalein was chosen from the chart in the lab manual.[3] As indicated in **Table 4,** from the three titrations the average molecular weight for the unknown was found to be 266 g/mol. The actual molecular weight for 3-nitrobenzoic acid is 267 g/mol.

Once the compound had been identified, the MSDS was retrieved and disposal and toxicity data were obtained. The compound has an LD50 (oral dose in a rat) of 1950 mg/kg, which indicates that the compound is not very toxic (compared to sodium cyanide, which has an LD50 of 6.4 mg/kg). However, it would be unacceptable to dispose of it into the drains. Our research team recommends that the material be placed in a sealed container and disposed of as an inert organic compound by a professional disposal company.

Synthesis

There were several possible syntheses of the compound. For example, benzoic acid can be nitrated to give the 3-nitrobenzoic acid, as shown in scheme 1.

There are several programs that can be used to draw chemical structures, and these structures were pasted into this document from such a drawing program. However, it is acceptable to draw structures by hand if you do not have access to such a program.

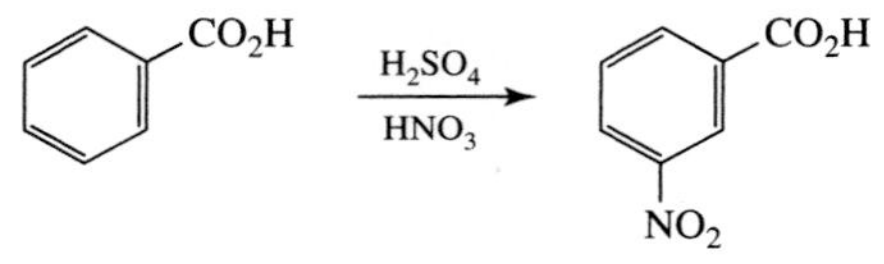

Scheme 1

However, when the safety aspects of this synthesis were investigated, we discovered that concentrated sulfuric and nitric acids would be needed and the laboratory instructor advised that this synthesis was not safe to perform in a general chemistry laboratory. Instead of a synthesis of the compound it was decided that a derivative of the acid should be made. That is, an ester of the acid would be synthesized as shown in scheme 2.

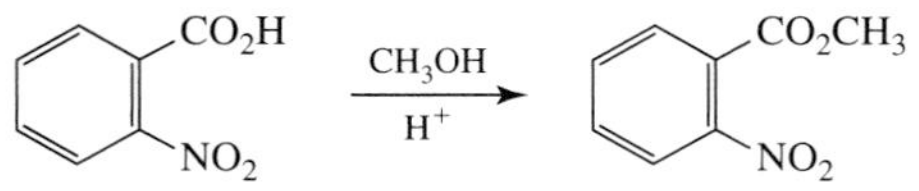

Scheme 2

The ester was prepared by dissolving 1.0 g of the nitro-acid in 25 mL of methanol and adding two drops of concentrated H_2SO_4. The reaction was monitored by TLC since the ester had a much higher retention factor, R_f, than the acid itself, because the ester is much less polar and thus less strongly adsorbed onto the plate. When the reaction was complete and no more starting material could be observed on TLC, it was poured into a solution of aqueous sodium bicarbonate. Since the starting material is an acid, it dissolved in the base, leaving the ester, which was insoluble in water and appeared as cream-colored crystals. The yield was 23%. One reason for the low yield might be that an incomplete reaction had occurred. In addition some mechanical losses might have occurred during filtering and transferring the material.

Note the explanation for the low yield here.

Since there was some extra time before the report was due, the group decided that it would be interesting to hydrolyze the ester back to the acid as shown in scheme 3 and see if the course of the reaction could be followed by some means. The solution of the acid in base was yellow, and it was possible that if the ester was treated with aqueous base the appearance of the yellow color could be monitored in some way.

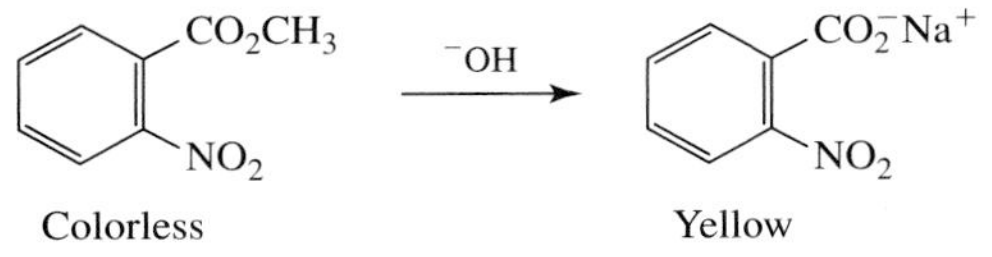

Scheme 3

The lab instructor suggested that the Spectronic 20 spectrophotometer could be used to detect the presence of the yellow anion, since the absorbance of a species in solution is directly related to the concentration of the species in solution.

In order to see if this approach was viable, several solutions of known concentration of the nitro-acid in dilute NaOH were prepared. Each absorbance reading was taken at 400 nm since the yellow color had a strong absorbance there. It is important to take absorbance readings at a wavelength where the absorbance is high in order to get the largest changes possible from reading to reading, thus leading to more accurate results. A graph of [acid] in 0.1 M NaOH versus absorbance gave a straight line (**Figure 2**). The next step was to begin a hydrolysis reaction and take out aliquots at predetermined time intervals, measure their absorbance, and then calculate the concentration of the hydrolyzed acid at that time. However this last step had to be omitted due to lack of time.

This group obviously had a good idea here and did not have time to finish up. It is OK to put experiments that are not finished in your report.

Conclusion

The conclusion is the place for a concise restatement of what you found during your experiments; it is not the place for expressions of satisfaction (or dissatisfaction) with what you learned.

The unknown compound was identified as 3-nitrobenzoic acid by a combination of tests of physical properties, such as melting point and solubility, chemical tests, and spectroscopy. The molecular weight was found by titration to be within 0.6% of the actual molecular weight. Since the MSDS indicated that the compound was relatively inert and nontoxic, it is recommended that the compound be disposed of by a professional disposal company, but no other special precautions need to be taken. Synthesis of a derivative of the acid led to the methyl ester as shown in scheme 2. A preliminary investigation into the hydrolysis of the ester was cut short through lack of time.

Experimental

Remember the experimental section should contain only the experimental details so that someone could repeat your work. It is safe to assume that the person who would repeat your work is fairly knowledgeable about chemistry. You do not need to put in long discussions of equipment or pictures, unless you have constructed a novel piece of equipment. Make sure you include observations of what actually happened in your experimental section.

Solubility Tests (Qualitative)

A small amount (about 0.1 g) of the unknown was added to 2 mL of the solvent in a test tube and shaken. The test tube was observed to see if any of the solid had dissolved. The solvents used were toluene, acetone, methanol, water, 0.1 M HCl, and 0.1 M NaOH.

The compound was most soluble in acetone, methanol, and NaOH. In NaOH the solution turned yellow, while in the other solvents the solution was a pale straw color.

Solubility Tests (Quantitative)

The compound was too soluble in acetone and methanol to test the quantitative solubility.

Solubility in NaOH

The unknown compound (0.205 g) was placed in a small Erlenmeyer flask and 5.00 mL of 0.1 M NaOH was added. The mixture was stirred and heated slightly for about 15 minutes. A watch glass was placed on top of the flask, and the solution was left to cool for one week. The resulting mixture was filtered and the solid was dried to constant weight. A 0.095-g-amount of solid was recovered, meaning that 0.110 g of solid was dissolved in a 5-mL solution, or the solubility is 22 g/L.

Chemical Tests

> *Do not forget to put your observations here. Do not simply copy out the procedure.*

Alcohols: Ceric Nitrate Test

Ceric nitrate (5 drops) was put into a porcelain test plate; a small amount of the unknown (dissolved in a few drops of acetone) was added. No color change was observed.

Aldehydes and Ketones: The 2,4-Dinitrophenylhydrazine Test
To a few drops of the DNP reagent in a porcelain test plate was added a drop of an ethanolic solution of the unknown. No color change was observed.

Phenols: The Ferric Ion Test
To a few drops of the test reagent in a porcelain test plate was added a drop of an ethanolic solution of the unknown. No color change was observed.

Chlorides: The Silver Nitrate Test
A few drops of silver nitrate solution were added to an aqueous solution of the unknown. No precipitate was observed.

Carbonates: The Acid Test
6 M HCl was added dropwise to an aqueous solution of the unknown. No effervescence was observed.

Sulfates: The Barium Chloride Test
A few drops of barium chloride solution were added to an aqueous solution of the unknown. No precipitate was observed.

Synthesis of the Methyl Ester of 3-Nitrobenzoic Acid
The nitro-acid (2.00 g, 1.20×10^{-2} mol) was dissolved in methanol (25 mL). Concentrated sulfuric acid (10 drops) was added, and the reaction was heated and stirred for 1 hour. Monitoring of the reaction by TLC indicated the presence of a new product. The reaction mixture was poured into 0.1 M NaOH, and the resulting cream-colored crystals were filtered, washed, and dried. Yield 0.50 g (23%), mp 74–76°C.

Preparation of Solutions for Spec 20 Analysis

The nitro-acid (0.167 g, 0.00100 mol) was placed in a 10-mL volumetric flask and dissolved in 0.100 M NaOH to make up the volume to 10 mL resulting in a 0.100 M solution. The solution was diluted by a factor of 10 twice, until a 0.00100 M solution was obtained, which had a transmittance of <5%. The solution was then further diluted by taking separate portions of 1, 2, 4, 6, and 8 mL and making the volume up to 10 mL with the 0.100 M NaOH. The absorbance of each sample was then recorded at 400 nm.

References

1. *Aldrich Catalog*, 2002–2003, p. 1039.

2. Tro, N. *Chemistry: A Molecular Approach,* Pearson: Upper Saddle River, NJ, **2008**.

3. Cooper, M. M. *Cooperative Chemistry Laboratories,* McGraw-Hill: New York, **2008**, p. 92.

Reporting Numerical Results

Significant Figures

When you carry out measurements in the lab and then do calculations based on those measurements, it is important to remember that the answers you get cannot have more significant figures than the initial data. The calculated answers you get cannot be more precise than the data you start with. You have probably had the rules for significant figures explained to you in lecture and in your textbook. However, it is in the lab where your use of significant figures actually becomes significant. For example, in a thermochemistry experiment you might measure a temperature change of 10.2°C. If you use that temperature change to calculate a heat change of 4.216953×10^2 kJ/mol, you are obviously overstepping the accuracy of your initial measurements.

Significant Figures Reminder

1. When adding or subtracting, use the smallest number of **decimal places** to decide how many significant figures your answer should have. For example,

   $2.02 \times 1056.1 = 1058.1$

2. When multiplying or dividing, use the number with the smallest number of **significant figures** to decide on your answer. For example,

   $2.02 \times 1056.1 = 2130$

   In the preceding case, the number of significant figures in the answer has now become ambiguous, which leads us to:

3. Use **scientific notation** for reporting data, so that the preceding answer would read $2.130 + 10^3$, thus removing the ambiguity in the significant figures.

Accuracy

The accuracy of a measurement concerns the agreement of the measurement with the true or correct value. The accuracy can only be known if the actual value is also known. Usually people will take several measurements and then average them. However, it is possible to report an accurate result from measurements that are not precise.

Precision

The precision of data is a measure of the agreement between two or more readings. Data points that are close together are precise, but they may not necessarily be accurate. The three targets below show the difference between precision and accuracy.

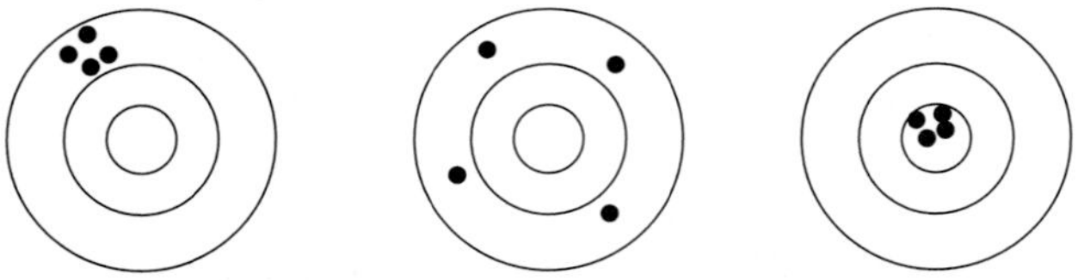

Precise but not accurate   Accurate but not precise   Accurate and precise

When you record data, in many cases you will not know what the true value of the data should be and therefore will not know if your data are accurate. If you take multiple readings of data, you will be able to ensure that your data are accurate, as long as the measurements are close together. However, precision does not necessarily mean your data are accurate since a systematic error could affect all your data in the same way. It is good practice to take multiple readings of data and to perform multiple runs of all experiments. **One of the hallmarks of good experimental work is that it is reproducible.**

Uncertainty

In any endeavor in the lab where you take measurements of any kind, it is inevitable that some degree of uncertainty will creep in. For example, if you find the mass of an object on a balance, the mass you get will depend on a number of things. It will depend on how carefully you take the measurement, how accurate your balance is, and how many readings of the mass you record.

The uncertainty, or reproducibility, of your result can be calculated fairly easily, and you should include the uncertainty of any measurement that you make.

There is an inherent uncertainty in the use of any measuring device or instrument; for example, balances are accurate to ± 0.001 g, and most volumetric glassware has the tolerance or range of uncertainty printed on it (or it can easily be looked up in a catalog). In addition to the absolute or inherent uncertainty of the equipment, other factors can add to the uncertainty of the measurement.

Imagine you have produced a new compound, have recrystallized it, and are now ready to find its mass. You might weigh it three times and find the following masses:

1. 1.216 g
2. 1.145 g
3. 1.137 g

Clearly the first mass is out of line with the following two masses. A common reason for this is that the solid might not be dry when first weighed. Weights (2) and (3) are not identical but are much closer. After another fifteen minutes or so of drying, weigh the material again.

The next weight is found to be

4. 1.141 g

Since the first weight is obviously not accurate, you should not use it in your report of the weight of the compound. It is perfectly acceptable to disregard data that are clearly in error as long as you have enough other readings to use.

You should record the average weight of readings (2), (3), and (4), which would be 1.141 g. However, the balance is accurate to ± 0.001 g, and this reading is 0.004 g away from the high and low readings. You could report the mass as 1.141 ± 0.004 g. This is the absolute uncertainty. Clearly this uncertainty is greater than that inherent in the balance itself, which means that there must be some other factor affecting the readings. Drafts might have caused a fluctuation of the reading, the balance needed servicing, or there was an operator error.

If you need to report data with a relative uncertainty, the measurement can be calculated by taking the absolute uncertainty, which in the preceding case is ± 0.004 g, and dividing it by the measurement itself.

$$\text{Relative uncertainty (RU)} = \frac{\text{absolute uncertainty}}{\text{actual measurement}} \times 100$$

In this case

$$\text{RU} = \frac{0.004}{1.141} \times 100 = 0.35\%$$

In this way you will have an idea of the percentage error of your measurements.

Graphing Data

There are many reasons for graphing data. One is that it can give you a pictorial representation of the data that might be easier to understand (this is why the media uses pie charts and bar graphs, which have a much greater visual impact than columns of numbers). However, for lab purposes, we usually graph data so that we can calculate some other quantity from the graph, and/or to see if we have a relationship between variables in an experiment. For example, if we have data from an experiment that looked at the concentration of a reactant with respect to time, there are a number of ways to analyze this data. The first way would be to report it in a table as shown here:

$[N_2O_5]$ (mol/L)	Time (s)
0.1000	0
0.0707	50
0.0500	100
0.0250	200
0.0125	300
0.00625	400

It is difficult to see the trends in data from simple tables like this. The obvious next step would be to graph the data. Since time is the variable that was set by the experimenter (the data were obviously recorded at set time intervals), time is the independent variable and is put on the *x* axis. The concentration is the variable that is measured. It is the dependent variable and goes on the *y* axis.

There are a number of types of plot you can use to visualize data. The easiest to start with is a simple pencil-and-paper plot.

As you prepare graphs and charts, remember the following points:

1. Use graph paper or prepare computer-generated graphs.
2. Label all tables, graphs, and charts with titles and a number so that the reader may easily see which graph or table you are referring to.
3. Tailor the size of the graph to fit your data.
4. Make the graph proportional. The intervals should be appropriate and clearly marked.
5. If there is more than one set of data in a chart or graph, make sure the reader can distinguish and identify the lines.
6. Include a legend with each graph to identify various parts.

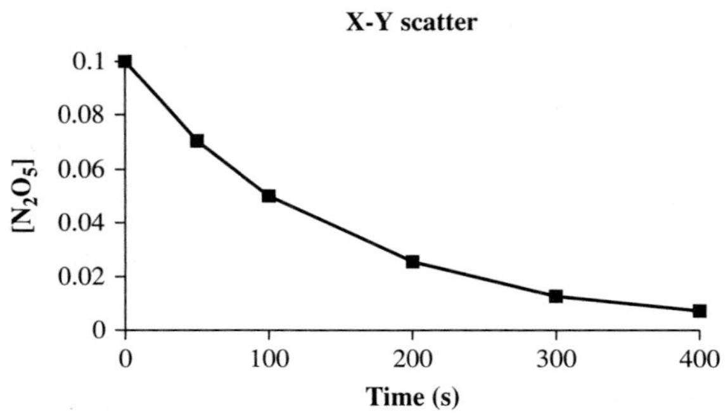

This computer-generated plot of the data shows how the concentration changes with time and is obviously not a straight line.

If you use computer-generated plots, it is very important to understand what you are doing, for example, how to get each variable on the right axis with an appropriate scale. It is possible to get into all kinds of trouble if you do not understand graphing programs. If in doubt, ask your instructor. Remember that a pencil-and-paper graph is always an option, at least to give you an idea of what the graph looks like. However, in many experiments you may need to generate a large number of graphs, and in the long run, it is much easier to do this on a computer.

The graph obtained from the data does not really give any more information about the reaction than the table of data. In order to get more information, there are a number of possibilities. For example, it is known that if the rate of reaction depends only on the concentration of one reactant, then an equation of the following type can be generated:

$$\ln [N_2O_5] = -kt + \ln [N_2O_5]_0$$

This is of the same format as the equation for a straight line:

$$y = mx + c$$

If we take the same original data and now plot the natural log of the concentration versus time, we will be able to see if our reaction behaves as we predicted. If you plot by hand, be sure to draw the best straight line. This is not a connect-the-dots exercise. You must draw the best straight line through the points, and this is often not an easy thing to do. It helps to use a transparent ruler. If you use a graphing program, it should draw the best straight line for you and then do any subsequent calculations or measurements you need.

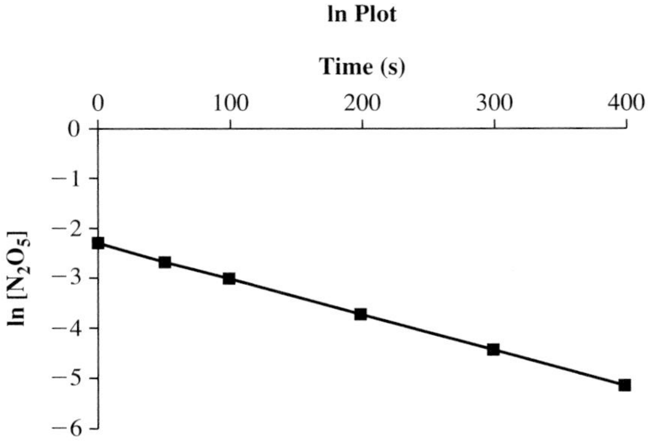

As predicted, a straight line is observed, and the slope of the line should be $-k$ (a value known as the rate constant for a reaction).

To find the slope of the line manually, take two points on the line and calculate the "rise over the run" or $\Delta y/\Delta x$. If you find the slope this way, be sure to **pick two points on the line that are not data points.** The whole idea of drawing a graph is to average out the errors in data collection. If you use actual data points, there would be no need to draw the graph; you could just use the raw data points and calculate the "change in *x* over the change in *y*." If you have used graphing software or a graphing calculator, it is a fairly easy matter to get the slope of the line using the built-in capabilities of the software.

The point is that when you decide to graph your data, you need to understand what you are doing and why you are doing it. Before you start you need to have a clear idea of what you are looking for in your graph and how you will go about finding it.

One last word: when you graph data, sometimes you will see a data point that looks to be in error. For example:

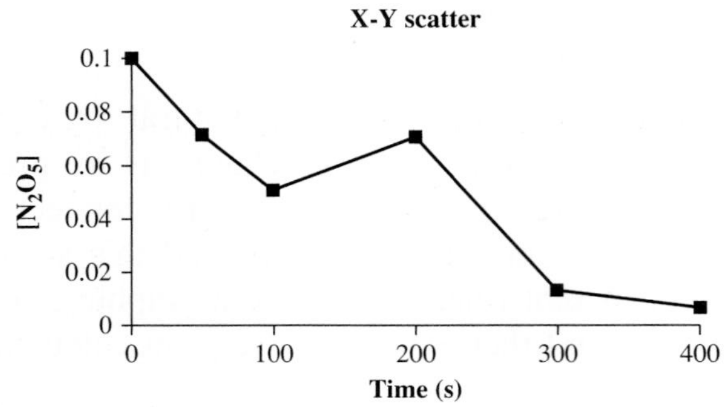

Concentration of N_2O_5 v time

One data point in this graph obviously has a problem! Ideally you would like to go ahead and rerun the experiment to see if the point was indeed an aberration, but sometimes this is impossible, and in this case the more data points you have, the more justified you would be in discounting that data point. If you only took three sets of data, you will have a hard time throwing one out. If you have twenty data points, the one you disregard will be less important.

Oral Presentations

During this course you will probably be required to give several presentations of different types. This guide is intended to give you some pointers so that your presentations will be more effective.

1. The presentation will be a group effort. Material should be divided so that each person has something interesting to talk about. In a group of four people an equitable distribution might be:

   Group leader. Explain why your group has chosen this project and the rationale behind your group's experimental strategy.

   Communicator. Continue discussion of experimental strategy; describe experimental setups.

   Record keeper. Describe what happened. Did you have to change your strategies as a result of your initial experiments?

   Counselor. Discuss what you learned. Did your experiments work as planned? What would you do to continue the project?

   This division is just a suggestion. Clearly the topics of each person's presentation will change from group to group and project to project.

2. When planning your talk, imagine that you will be teaching the material you discuss to your audience rather than simply reporting on it. Your points will come across better if you make an effort to explain and reach out to your audience.

3. **Use visual aids.** This can be as simple as writing on the chalkboard or using prepared charts and graphs. Remember that a picture is worth a thousand words.

4. A well-organized presentation is more likely to be remembered, and the presenter of a well-organized presentation is more likely to be remembered favorably. If your listeners can see how your points relate to each other and to an overriding theme, those points will carry more meaning.

5. Remember that each group in your lab section will be presenting their work, which means that some of the material will be very similar to yours. **It is your job to make the presentation interesting to the audience.** We encourage you to think of alternate ways to present your oral report. In the past many groups have developed different scenarios that they used to present the information. **You will not be penalized for making your presentation more interesting.**

6. Ask questions. It is highly probable that some of your grade for the oral presentation will be based on your interactions with other groups who are presenting their projects.

Section 2

Laboratory Equipment

CONTAINERS

Test Tubes

Test tube

There are two main types of test tubes: those made out of regular glass and those made out of Pyrex. Most of the test tubes in your lab drawer will be glass test tubes. These are fine but do not stand up to excessive and rapid heating and cooling as well as Pyrex test tubes do; however, they are cheaper. Test tubes are usually measured by their outside diameter and their length. For example, you may have 15 mm × 125 mm and 13 mm × 100 mm test tubes in your drawer. Test tubes may be used as containers for solids or liquids. They can be used as containers for quick tests for properties such as solubility and effect of heat. Often they are used to carry out reactions on a small scale and can also be used as centrifuge tubes when a separation of solid from liquid is necessary.

Beakers

Beaker

Beakers are larger containers than test tubes, usually with a pouring spout. The ones in your drawer are made out of Pyrex glass and are safe to use for heating materials. Beakers can come in all sizes from 5 mL to several liters. They can be used for carrying out reactions, heating solutions, and for conducting water baths or air baths. Many beakers have graduations on the side and can be used as approximate measuring devices for liquids. If a more accurate measurement is required, however, you should use a measuring cylinder or a volumetric pipet.

Erlenmeyer Flasks

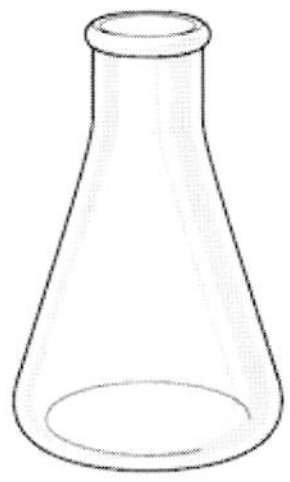

Erlenmeyer flask

These conical-shaped flasks are usually made out of Pyrex glass and are safe for heating solutions. The fact that the mouth of the flask is narrower than the bottom means that these flasks are particularly useful for containing volatile solvents. The small neck of the flask prevents solvents from escaping the flask. Erlenmeyer flasks are often used for recrystallizations, for reaction vessels, and for heating volatile solvents.

Microscale Cell Wells

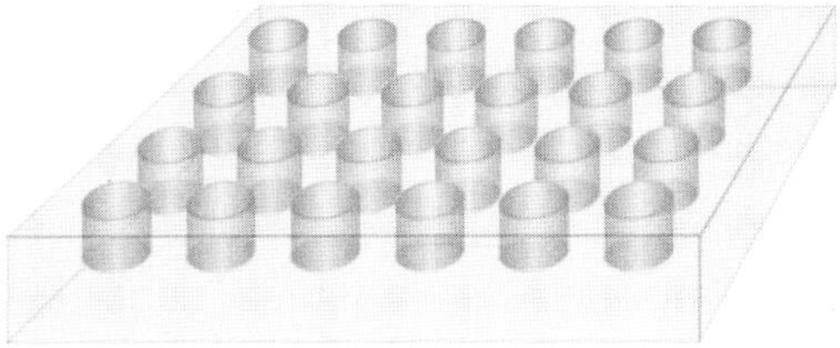

These plastic plates come in several sizes: from 6 to 96 wells per plate. The wells are labeled by letters for the horizontal rows and by numbers for the vertical rows, allowing you to keep track of which solutions you have in each well as long as you remember to record them. For instance, well

C4 would be the fourth well on row C. These cell well plates can be used for small-scale reactions and observation of properties such as solubilities. However, the wells cannot be heated with a Bunsen burner because they will melt, and they cannot be used for organic solvents (i.e., anything that is not water) because the plastic plates may dissolve.

Evaporating Dish

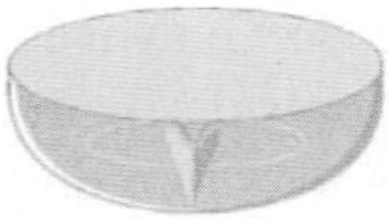

The evaporating dish is a shallow porcelain bowl usually used as a vessel to evaporate solutions to dryness. Once the solvent has been removed, the evaporating dish can be heated strongly.

Crucible and Lid

The crucible is a lidded porcelain container used to heat solids. Reactions of solids that are brought about by strong heating are often performed in a crucible.

Watch Glass

Watch glass

The watch glass is a very shallow glass bowl often used to allow crystals to dry after they have been filtered. It is not made of Pyrex and should not be heated to extremes.

Ignition Tube

An ignition tube is a large rimless Pyrex test tube. It can be used simply as a large test tube or as a reaction vessel, for example, in the production of gases.

MEASURING DEVICES

Measuring Liquids by Volume

Liquids can be measured either by weight or by volume. It is often more convenient to measure the volume of a liquid and then convert it to mass, if the density is known (remember, density = mass/volume).

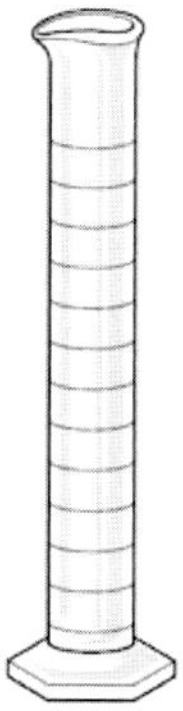

Graduated cylinder

Graduated Cylinders

A graduated cylinder can be used to measure the volume of a liquid. Graduated cylinders range from 5 mL to several liters, but you should be aware that the accuracy of a graduated cylinder is not as great as volumetric glassware (see below). Measuring with a graduated cylinder is fine for most purposes, unless you are doing accurate volumetric measurements such as titrations.

Volumetric Glassware

Volumetric glassware is used when great accuracy is required, for example, during titrations or when making up standard solutions of known volume.

Pipets

Calibrated glass pipet

Most volumetric pipets are made to measure one volume only, and the most common ones are 10, 25, and 50 mL. They are used to measure a specific volume of liquid and will usually have the accuracy engraved on the pipet. Pipets must be used with some sort of pipet filler for suction. **Never pipet by mouth.**

Burets

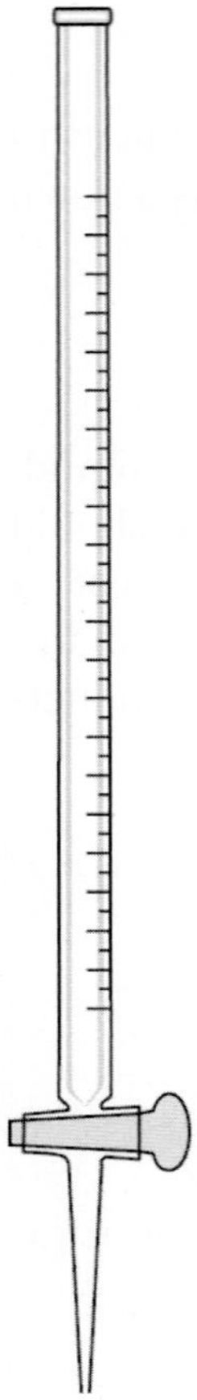

A buret is used for measuring varying volumes of liquids and for delivering volumes of liquids accurately. Usually the buret is calibrated in divisions of 0.1 mL and can be read to an accuracy of 0.01 mL. To prepare a buret for use, wash it and rinse it thoroughly with tap water. Then rinse with two 5-mL portions of your solution. Fill the buret above the zero mark with the solution; then open the stopcock and allow the solution to drain to, or just below, the zero mark, making sure the tip of the buret is filled. Read the bottom of the meniscus for your initial buret reading.

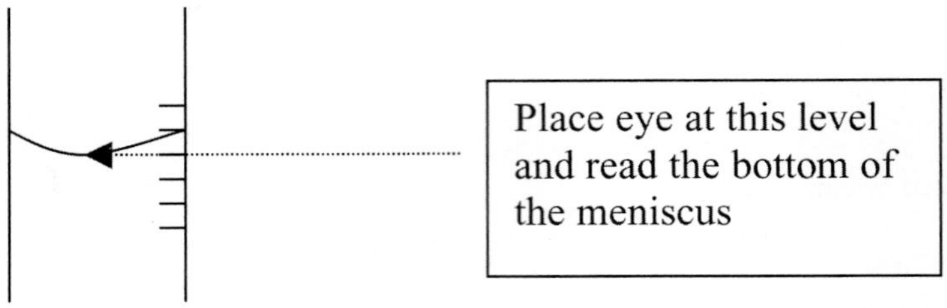
Place eye at this level and read the bottom of the meniscus

Volumetric Flasks

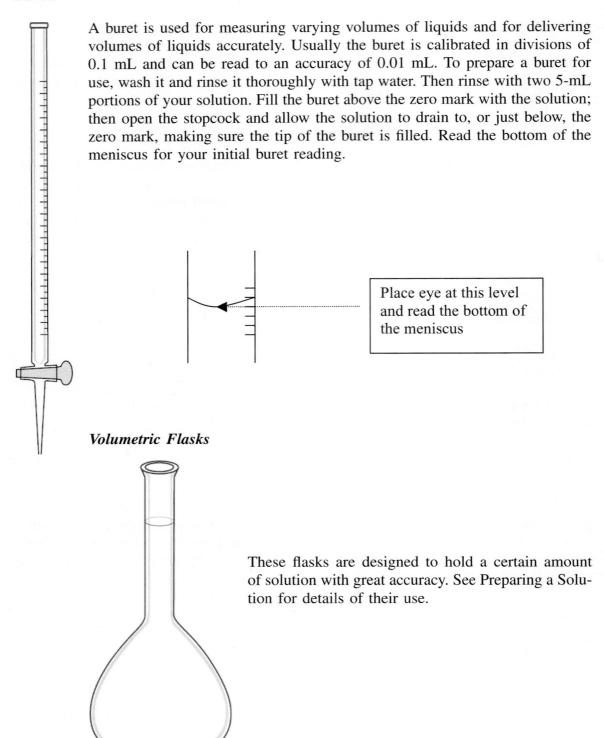

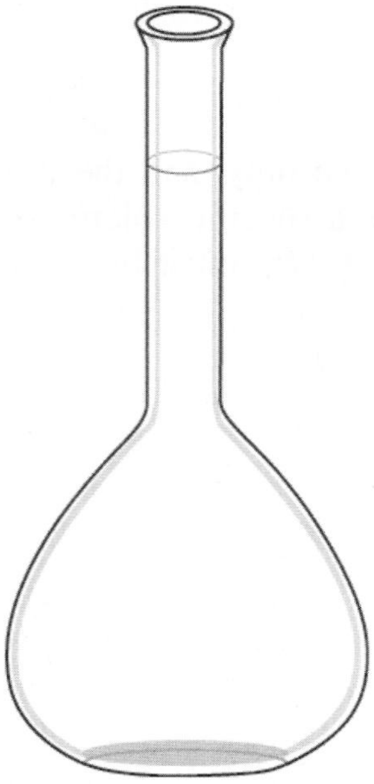

These flasks are designed to hold a certain amount of solution with great accuracy. See Preparing a Solution for details of their use.

Measuring Solids and Liquids by Mass

Amounts of solids are usually measured by weighing them.

Laboratory Balance

A laboratory balance is used to obtain the mass of various objects. There are several different types of balances, but probably in this course you will be using a digital, top-loading balance. These balances are accurate to 0.001 g and simple to use, but they are delicate and expensive. Please treat them with care and respect, and leave them clean. If you spill anything solid on the balance, use the attached brush to clean it. If you spill a liquid, notify the stockroom personnel immediately.

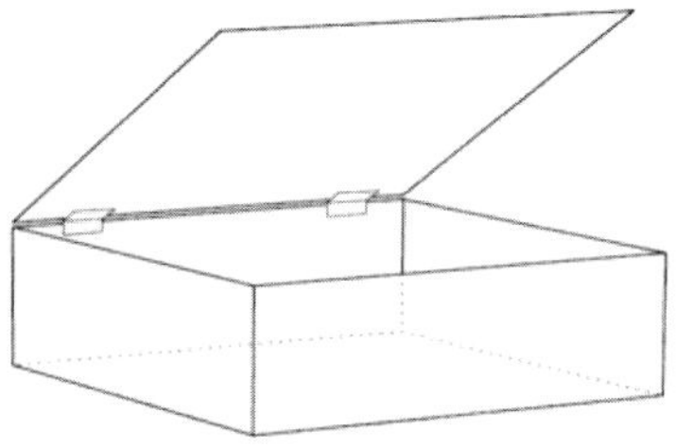

Note: The balances are fitted with draft shields to remove fluctuations of the readings. Please do not remove them.

Use of Balance

1. The balance has only one control: the control bar. If the balance is off, switch on the display by pressing the control bar briefly. The balance will eventually read 0.000 g.
2. To weigh an item, place it on the pan, wait till the balance is stable, and read the result.
3. Most items you will need to weigh are chemicals. DO NOT PLACE CHEMICALS DIRECTLY ON THE BALANCE. You may weigh your material on a piece of the provided weighing paper, or in a beaker or watch glass. The balance will automatically subtract the weight of your container if you follow this procedure:

   a. Place the empty container on the balance pan; wait until the readout is stable.

   b. Press the control bar briefly.

   c. The balance will read 0.000 g, and you may now place your item to be weighed in the container and read off its weight.

   Note: This procedure is called taring; the container is said to be tared when its mass is zeroed out of the reading. The balance is accurate to 0.001 g ± 1 mg. Data and calculations from the balances should normally be recorded to this accuracy and no more.

TRANSFER DEVICES

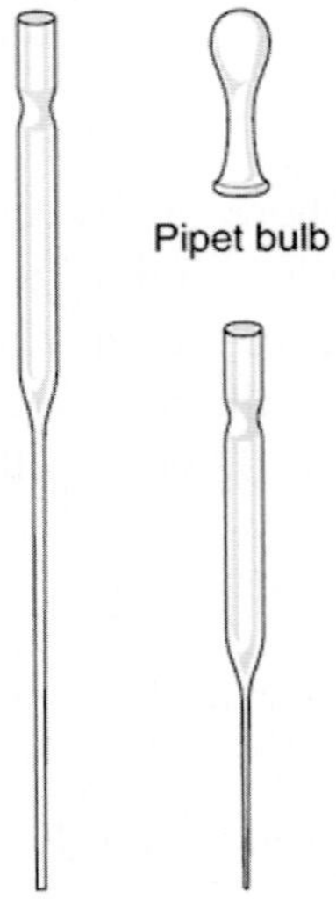

Pasteur pipets

Droppers and Pipets

Several different types of pipets can be used for transferring volumes of solutions or adding solutions dropwise to a reaction mixture: Glass pipets, such as Pasteur pipets or medicine droppers, usually have a rubber bulb that attaches to the top and acts as the suction device to draw liquid up into the pipet. Plastic pipets, sometimes called Beral pipets, have the bulb built in as an integral part of the pipet. These plastic pipets can be heated and the tips drawn out so that they deliver a very small amount of liquid. (See **Microscale Techniques** in Section 3.)

Funnels

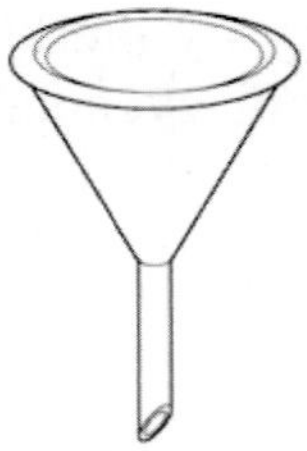

Conical funnel (glass)

There are two main types of funnels. A simple **gravity funnel** may be used as a guiding device to help when pouring liquid from one container to another. However, with the aid of a filter paper, it can be used as a separation device to separate liquids from solids.

Büchner funnel (porcelain)

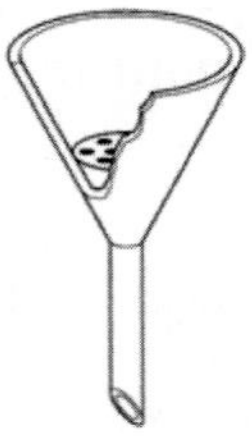

Hirsch funnel (porcelain)

A **Büchner funnel** is used to separate a solid from a liquid. A piece of filter paper is placed over the holes in the funnel. In general, a Büchner funnel and suction are used when the solid is required, and gravity filtration is used when the liquid is required.

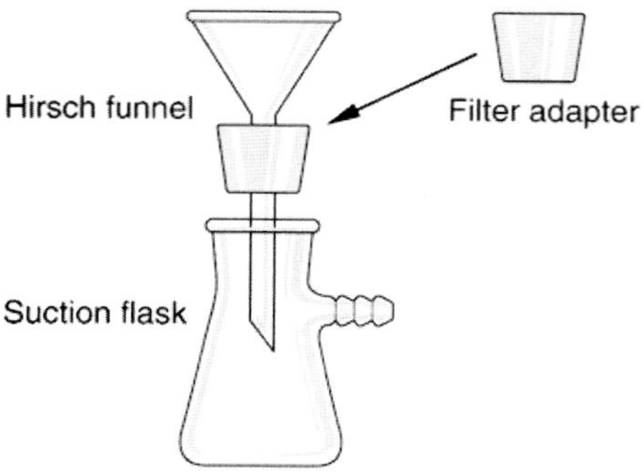

A **Hirsch funnel** is a conical-shaped suction funnel and is used for small-scale filtrations as shown in the picture.

Tongs

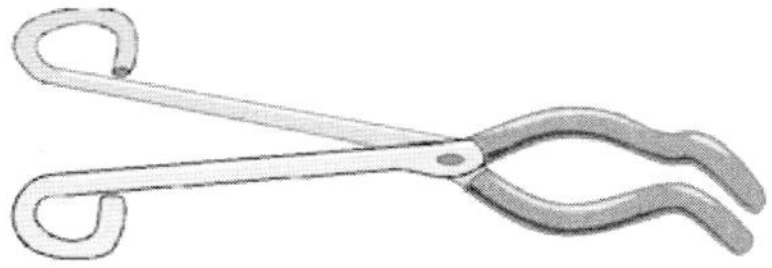

Tongs can be used to transfer hot objects from one place to another.

Test-Tube Holder

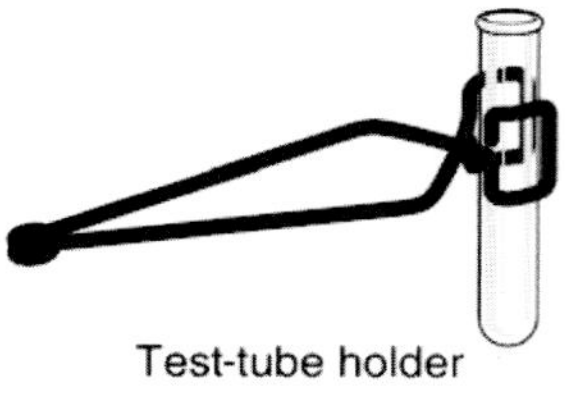

This device is used to hold a test tube when it is being heated.

Forceps

Forceps are used to pick up relatively small objects.

Spatulas

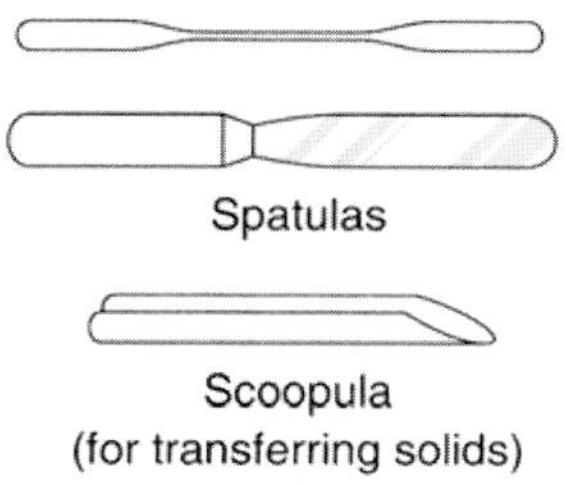

Spatulas can be used to transfer solids from one place to another. They come in a number of different forms and sizes.

SUPPORT DEVICES

Metal Ring

Metal ring

A metal ring attaches to the stand on your benchtop. It can serve as a support for pipestem triangles and wire gauzes.

Pipestem Triangle

The pipestem triangle is used to support a crucible while it is being heated over a Bunsen burner. It is very heat resistant and can be used at high temperatures.

Wire Gauze

Wire gauzes are used to support larger objects, such as beakers and evaporating dishes, while they are being heated. Like the pipestem, the gauze can be used at high temperatures.

Clamps and Clamp Holders/Universal Clamp

Clamps are used to hold equipment steady; they are usually held by a clamp holder that attaches to the stand on the bench top.

A universal clamp has both the clamp and holder in one piece and performs the same function.

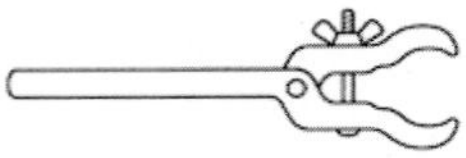

Clamp

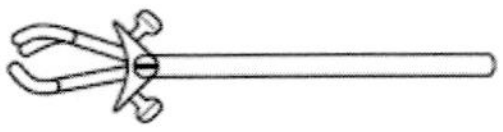

Three-finger clamp

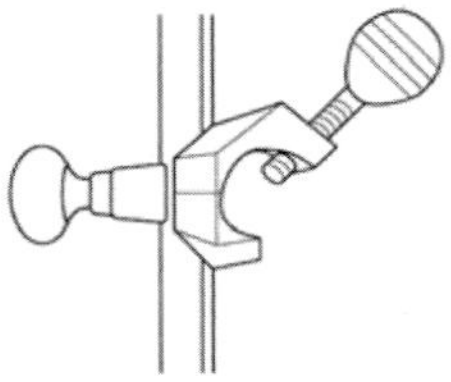

Clamp holder

HEATING DEVICES

Bunsen Burner

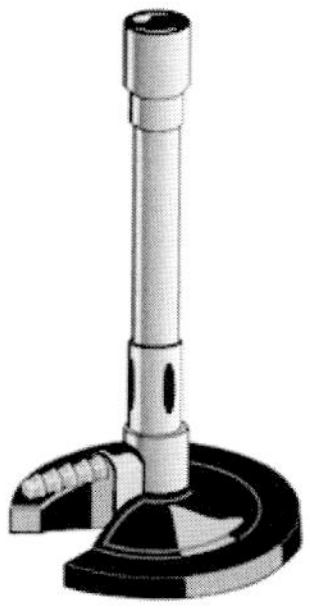

The Bunsen burner is used frequently in the laboratory as a source of heat when no flammable material is present. The burner allows a useful controllable flame. In order to achieve the best flame, the gas inlet valve and the air vents on the barrel should be properly adjusted.

Your burner should be equipped with a piece of rubber tubing. Attach one end to the sidearm and the other end to the gas outlet on your lab bench. Turn the barrel of your burner completely down, and make sure the gas inlet valve is open about halfway. Open the gas outlet on your bench completely, and light the burner with a match. Adjust the gas inlet valve so that the flame is about 4 inches high. The flame should be luminous and yellow, which is caused by incomplete combustion of the gas because of insufficient oxygen. The carbon particles that are formed become incandescent.

This luminous flame is not very hot (comparatively!) and may be used for mild heating. However, this flame does leave a film of carbon particles (soot) on anything it contacts. You can observe this phenomenon by holding your evaporating dish in the flame for a minute or so. (Use tongs to hold the dish!)

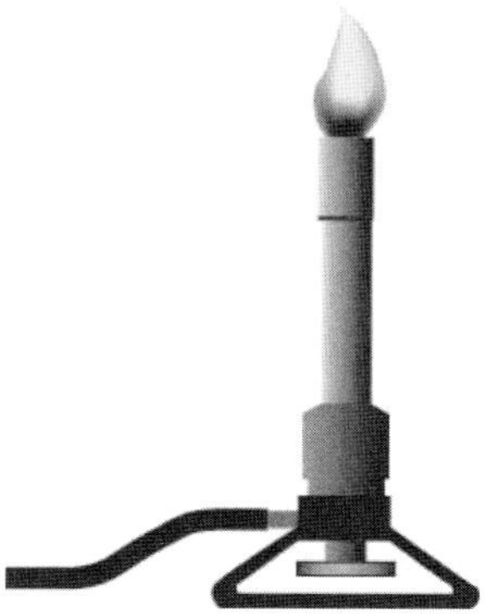

In order to make the flame hotter you need to admit more oxygen into the mixture to be ignited. Rotate the barrel by holding it near the air holes, until the flame becomes light blue and nonluminous. Too much air may blow out the burner or cool the flame. Two zones will be apparent in the flame, the outer zone and the inner zone. This flame should be used for strong heating. In order to check that all the gas is now being combusted, hold your dish in the flame and see if any more soot is deposited. If soot is deposited, combustion is incomplete. Test the various zones of the flame with an unlighted match or a wire gauze to see which is the hottest.

Heating Bath

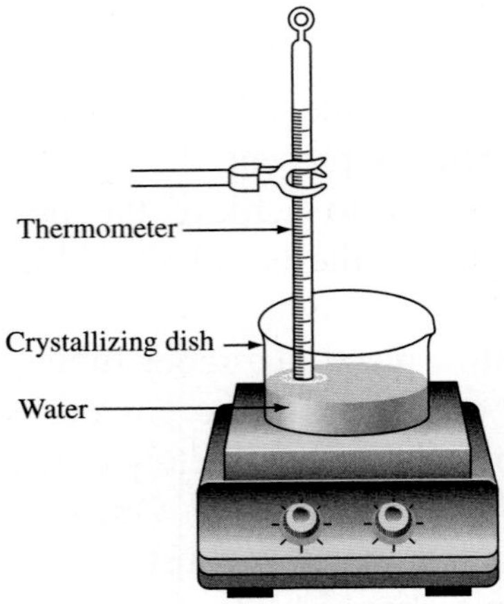

As previously stated, a Bunsen burner may reach very high temperatures. However, it is often necessary to heat reactions or solutions in a more controlled manner so that the temperature does not exceed a certain value. For example, many reaction mixtures might decompose when heated to a very high temperature. One way to avoid this problem is to use a water bath. In a water bath, the Bunsen burner can be used to heat a beaker of water, which in turn is used to heat the desired material. The material to be heated is placed in the beaker, and as long as there is water in the water bath the temperature will not get higher than 100°C (the temperature of boiling water). There are other possibilities for heating baths such as sand baths and oil baths, but these are more often used with electrical hot plates as the source of heat as shown in the illustration.

Section 3

Laboratory Techniques

PREPARING AN EXPERIMENT

Let's say you have decided on an initial course of action. Before you begin you should be able to answer the following questions:

1. Why are you doing this particular experiment? This will, of course, depend upon your project, and each project will consist of many different experiments. However, you should be able to give an answer to this question at each stage of your project development.
2. What equipment will you need to carry out the experiment? Is it available in the lab? In the building? Can your group make the piece of equipment you need?
3. Which reagents will you need? Are the concentrations of solutions appropriate for the experiment? Will your instructor or the stockroom personnel be able to find the right chemicals for you by next week? Do you have an alternate plan if all your reagents cannot be supplied in time?
4. **Have you looked up all the reagents you will use and noted any safety precautions and special disposal instructions?** Material safety data sheets (MSDSs) are available in the laboratory for most compounds. If there is not one for a particular compound that you are planning on using, you should generate one by using the on-line information system and add it to the laboratory stock.

The following are a few suggestions that you should bear in mind before you begin your practical work.

1. If you are planning to do any kind of reaction, especially if you have not performed this reaction before, you should always begin with a trial reaction on a small scale. A good way to begin is to use the microscale cell wells as reaction vessels and the disposable pipets to deliver liquids and solutions. (If you are not familiar with microscale chemistry, there is a section devoted to it in this lab manual, your lab instructor may demonstrate it for you, or there is information on-line.)
2. If your reaction seems to be successful on the microscale, then you can scale up the reaction, but you should be aware that it is often the case that things go wrong when a reaction is scaled up. For example, if the reaction is exothermic (gives out heat), then this effect will seem much more marked on the larger scale, and the reaction vessel may get quite hot. **You should not attempt to deal with quantities of compounds greater than 2.0 g unless explicitly told to do so by your instructor.** Be sure to take all appropriate safety precautions. All large-scale reactions should be done in your hoods.
3. Make sure you have all the reagents and equipment on hand before you start your experiment.

4. Make a list of everything you will need that does not seem to be readily available in the lab and give it to your TA the week before you plan to do the experiment.

5. Make sure that you apportion the work equally. Usually no more than two people would need to collaborate on any one experiment. It is a waste of time and resources for one person to perform the experiment and the other three group members to watch. There is plenty of time during lab for you to pool your ideas and results and to discuss their implications.

DEALING WITH UNKNOWN COMPOUNDS

Preliminary Tests

It is important when dealing with unknown compounds to proceed in an organized manner. You can save yourself a good deal of time and learn a fair amount about your compound by performing a few preliminary tests, as long as you understand the reasons for performing the tests and the implications of their results. When performing these tests, it is important to conserve your material; that is, you should use microscale quantities whenever possible.

Smell

If your compound has a distinct smell, you can conclude that it is not ionic. (Why not?) To see if a compound has an odor, gently waft the odor toward yourself with your hand. Do not smell directly; the odor may be overpowering.

Physical State

The physical state of your compound can give an indication of the nature of the compound. For example, if your compound is a liquid, it will not be ionic. (Why not?)

If your compound is a **solid**, a thorough examination under a magnifying glass may reveal whether the material is homogeneous. Crystal shape may also help in classification of the compound. General chemistry labs are not equipped to do melting points on high melting solids such as ionic compounds; however, you can investigate the effect of heat by carefully heating a small sample on a spatula. Subsequent examination through a magnifying glass may reveal changes in the crystalline state. As an example, many compounds decompose on heating. In general you will not be able to observe melting points for simple ionic compounds (the melting points are too high), but covalent compounds will melt or decompose when heated, as will more complex ionic compounds such as ammonium chloride.

If the compound is a **liquid**, it may be possible to obtain its boiling point (see the later section **Boiling Points/Melting Points**). If not, you can observe how a very small amount behaves when heated by carefully heating a small amount on a spatula. If the liquid is flammable, you should observe what type of flame is produced. **Note: Use very small samples for this test**.

SOLUBILITY TESTS

Qualitative Testing

A quick qualitative test for solubility can give you a fair amount of information about a compound. The first rule of thumb is: "like dissolves like." If your compound dissolves in water, it is probably a polar or ionic compound. In general, the solubility of ionic compounds decreases as the charge on the ions increases. If the compound dissolves in a nonpolar organic solvent like hexane or toluene, then the compound is probably non-polar. Solubility in acid or base solution can also give some indication of the properties of the compound.

Method

Take a small amount of the unknown (the size of a grain of rice), and place it in one of the compartments of a cell well plate. Repeat with as many samples as there are solvents you want to test. Fill each well up with a different liquid and stir. Observe carefully to see if any of the compound goes into solution.

Suggested Liquids	Inference if Soluble
Water (test with pH paper)	Polar or ionic compound
If not soluble in water, try 1 M NaOH in another well.	Probably an organic acid
If not soluble in water, try 1 M HCl in another well.	Probably an organic base
Toluene	Nonpolar (organic)
Acetone	Polar or nonpolar, probably not ionic

If your compound appears noticeably soluble in any of the liquids, you should do a quantitative solubility test.

Quantitative Testing

In order to accurately measure how much compound will dissolve in a given volume of solution, you must decide how much of your solute and solvent you will use. Since it will be impossible to weigh the exact amount of solute that will dissolve in a volume of solvent, you should accurately weigh out more solute than you need and put it in an **Erlenmeyer flask**. Then using a **volumetric pipet** add a measured volume of liquid, and heat the solution with stirring until no more solute seems to go into solution. (It is necessary to heat the solution because solids often take long periods of time to dissolve at room temperature.) Place a watch glass on top of the Erlenmeyer, and let the solution cool until any precipitation is complete, probably until the next lab period. Filter off the remaining solid, allow it to dry, and weigh it. You will then be able to calculate the mass of solute in solution and thus be able to calculate the solubility of the compound in grams per liter.

General Solubility Guidelines of Ionic Compounds in Water

Soluble Compounds	Exceptions
Almost all sodium (Na^+), potassium (K^+), and ammonium (NH_4^+) salts	
All chlorides (Cl^-), bromides (Br^-), iodides (I^-)	Ag^+, Hg_2^{2+}, Pb^{2+}
All fluorides (F^-)	Mg^{2+}, Ca^{2+}, Sr^{2+}, Ba^{2+}, Pb^{2+}
All nitrates (NO_3^-), chlorates (ClO_3^-), perchlorates (ClO_4^-), and acetates (CH_3COO^-)	Acetates of Ag^+ and Hg^{2+} are only moderately soluble.
All sulfates (SO_4^{2-})	Sr^{2+}, Ba^{2+}, Pb^{2+} (Ca^{2+} and Ag^+ moderately soluble)

Poorly Soluble Salts	Exceptions
All carbonates (CO_3^{2-}), phosphates (PO_4^{3-}), oxalates ($C_2O_4^{2-}$), and chromates (CrO_4^{2-})	Na^+, K^+, NH_4^+
All sulfides (S^{2-})	Alkali and alkaline earth metal ions and NH_4^+
All hydroxides (OH^-) and oxides	Alkali metals (those of Ca, Sr, and Ba only moderately soluble)

ANALYSIS OF ANIONS

All chemical tests should be done with the compound **in solution** if at all possible. It is not necessary to make up accurate solutions; a small amount on the end of the spatula in 1–2 mL of water will be sufficient. Many tests for anions rely on precipitating an insoluble salt of that anion. So, for example, the presence of a halide ion can be confirmed by adding a source of silver ions [$AgNO_3(aq)$] resulting in the precipitation of AgCl. Sulfate ion can be confirmed by adding a source of soluble barium ions ($BaCl_2$), resulting in the precipitation of $BaSO_4$.

It is good practice to run these tests on samples of known compounds first, before you test your unknown, so that you will know what a positive test looks like.

1. **Chloride** (in the absence of sulfate, bromide, and iodide). Place 1 mL of the unknown solution in a test tube, and add 1 mL of 6 M HNO_3 and 1 mL of $AgNO_3$ solution. The formation of a white precipitate (AgCl) shows the presence of the chloride ion, Cl^-. This test also gives positive results for any halogen or sulfate, since those silver salts are also insoluble.

   $Cl^-(aq) + AgNO_3(aq) \rightarrow AgCl(s) + NO_3^-(aq)$

2. **Sulfate**. Place 1 mL of the unknown solution in a test tube, and then add 1 mL of 6 M HCl and 1 mL of $BaCl_2$ solution. The formation of a white precipitate ($BaSO_4$) proves the presence of the sulfate ion, SO_4^{-2}.

   $SO_4^{-2}(aq) + BaCl_2(aq) \rightarrow BaSO_4(s) + 2Cl^-(aq)$

3. **Iodide or bromide**. Place 1 mL of the unknown solution in a test tube. Now add 1 mL of CCl_4 and 1 mL of freshly prepared chlorine water. Shake vigorously. A purple color in the CCl_4 layer confirms the original presence of the iodide ion, I^-, and yellow indicates bromide Br^-.

   $2I^-(aq) + Cl_2(aq) \rightarrow 2Cl^-(aq) + I_2$ (purple)

4. **Nitrate** (in the absence of iodide). Place 1 mL of the unknown solution in a test tube, and cautiously add 3 mL concentrated H_2SO_4. Mix thoroughly. After cooling the mixture, incline the tube at a 45-degree angle, and very carefully pour down the inner side of the tube 2 mL of $FeSO_4$ solution so that the latter floats on the top of the heavier liquid. A brown ring at the junction of the two liquids confirms the presence of the nitrate ion, NO_3^-.

5. **Carbonate**. Place 1 mL of unknown solution in a test tube and add 6 M HCl drop by drop. After the addition of each drop, look for effervescing in the tube. If your solution does effervesce, this proves the presence of the carbonate ion, CO_3^{2-}.

   $CO_3^{2-}(aq) + HCl(aq) \rightarrow CO_2(g) + H_2O + 2Cl^-(aq)$

6. **Acetate**. Place 2 mL of your unknown solution in a test tube, and add 1 drop of concentrated sulfuric acid. Add 1 mL of ethanol, and heat the solution for a few minutes in a water bath. The fruity smell of ethyl acetate indicates the presence of acetate.

   $CH_3COO^-(aq) + CH_3CH_2OH(aq) \rightarrow CH_3COOCH_2CH_3$ (fruity smell)

ANALYSIS OF CATIONS

Cations can be identified by precipitation of an insoluble salt. Add a solution containing the sodium salt of an anion that will form a precipitate with the cation you are testing for to a solution of the unknown compound. For example, if you suspect the compound contains calcium, you could add a solution of sodium carbonate to the unknown solution and see if calcium carbonate precipitates.

$$Ca^{2+}(aq) + Na_2CO_3(aq) \rightarrow CaCO_3(s) + 2Na^{+}(aq)$$

Unfortunately tests like this will not differentiate between the cations that form insoluble carbonates—so the cation tests are less specific than the anions.

It is good practice to run these tests on samples of known compounds first, before you test your unknown, so that you will know what a positive test looks like.

Test for Ammonium (NH_4^+)

To 1 mL of solution add 1 mL 6 M NaOH, and smell the resulting mixture by wafting the air above the test tube gently toward your nose. The smell of ammonia indicates the presence of ammonium ion. Ammonia may also be detected by holding a piece of moist pH paper near the neck of the test tube—ammonia will dissolve in the water on the pH paper, which will indicate a pH greater than 7.

Flame Tests

Cations may also be identified by flame tests. The technique of vaporizing a sample and noting the color imparted to the flame is called flame testing. Flame coloration is caused by the excitation of an electron to a higher energy level followed by a subsequent decay of the electron to a lower energy level with the emission of light. Some elements that impart color are listed in the following table.

To perform the test, obtain a clean nichrome wire and heat it strongly in a Bunsen flame to clean it. Then dip the wire into your unknown sample and reheat. Certain colors are more intense and brilliant than others and, unfortunately, obscure those of less brilliance and intensity. For example, even traces of sodium ions will give the yellow color of sodium, which will obscure practically all others. Since a mixture of ions is not useful in identification, one must employ some method to remove interference. One technique is to use colored filters. Sodium interference may be removed by using a blue cobalt-glass filter, which will absorb the yellow light of sodium and will transmit blue and violet light.

Flame Coloration

Element	Color	Intensity
Barium	Pale green	Low
Calcium	Red	Medium
Potassium	Pale violet	Low
Sodium	Bright orange/yellow	High
Strontium	Crimson red	Medium
Lithium	Carmine red	Medium
Copper	Blue-green	Medium
Magnesium	None	None

MICROSCALE TECHNIQUES

When performing an experiment for the first time, it is always a good idea to do the experiment on a small scale. This will give you some idea of what to expect, and should anything unforeseen happen, i.e., an accident, it is much easier to contain small amounts of chemicals. Small-scale experiments are safer, produce less waste, and are therefore preferred if possible. Obviously you will not be able to do all your work on the microscale, since it is very difficult to do quantitative measurements on a small scale. Microscale work is very useful when you want to see what will happen, for example, to see if a precipitate will form or a gas will evolve.

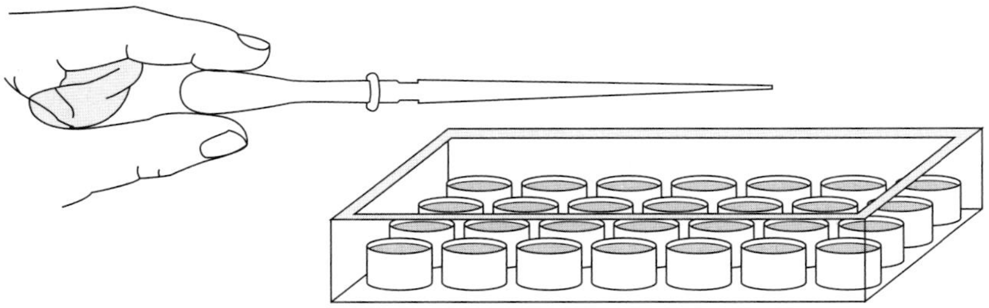

The pieces of equipment used for most microscale work are the cell well plates and the plastic pipets discussed in the laboratory equipment section. For semi-quantitative purposes the pipets can be used to deliver a constant volume by marking a point on the stem of the pipet. If the pipet is then filled to this level each time, even though you do not know what the volume contained in the stem is, you will be able to control the relative volume of each solution that you mix (i.e., 2 pipet volumes of NaOH per 1 pipet volume of H_2O).

One drawback to using microscale techniques such as these is that it is difficult to heat the test solutions. If you find that this will be necessary, you will have to scale up your reaction, perhaps to a test-tube scale, where you will be dealing with 1–2 mL instead of drops of solution. Material can be heated in a test tube either by direct heat in a Bunsen flame, or in a water bath. If you heat material in a flame, you should hold the test tube with a test-tube holder and pass the test tube in and out of the flame. If you allow the test tube to remain in the flame, the material in the tube will be heated too vigorously, which may result in charring of solids or bumping of liquids, causing them to fly out of the test tube. In any case, you should always heat test tubes while pointing the opening away from yourself or any other person working in the lab.

SOLUTION TECHNIQUES

To Make up a Solution of Known Concentration

1. Decide on the concentration you need. For example, if you wanted to make up a 0.1 M solution of NaCl:

2. Calculate the molecular weight of your compound (NaCl = 59.45 g/mol)

3. Decide how much solution you are going to make, and obtain a volumetric flask for that volume. Volumetric flasks are routinely available in sizes of 25, 50, 100, 250, and 500 mL and 1 L. If you are going to make up 250 mL of a 0.100 M solution of NaCl, you would need

$$(250 \text{ mL}) \times \left(\frac{0.100 \text{ mol}}{1\text{L}} \text{ NaCl}\right) \times \left(\frac{1 \text{ L}}{1000 \text{ mL}}\right) \times \left(\frac{59.5 \text{ g NaCl}}{1 \text{ mol NaCl}}\right) = 1.488 \text{ g NaCl}$$

   When you weigh your NaCl, weigh it as accurately as possible. The whole point of making up a solution this way is to know the concentration as accurately as possible.

4. Weigh out accurately on the most sensitive balance you can find, about 1.5 g NaCl on a piece of smooth weighing paper. In general chemistry labs you will find the balances weigh to 0.001 g. You will find it very difficult to weigh out exactly 1.488 g NaCl, and you should not try to do this. If you know exactly how much is there, you will be able to calculate the concentration of your solution. It is far better to weigh out an amount close to 1.488 g (say, 1.556 g) than to spend an excessive amount of time trying to get exactly 1.488 g.

   If your compound is a liquid, weigh it in a small Erlenmeyer flask or beaker.

5. Place the weighed compound into a beaker by making a kind of funnel with your weighing paper and directing the solid into the container. Then add about half the solvent you need to make up the volume. Swirl the beaker to dissolve the solute, and then pour the solution carefully into the volumetric flask. Wash out the beaker with a small portion of fresh solvent, and add the washings to the flask; repeat several times. Then make up the solution to the mark on the flask by adding solvent carefully from a wash bottle. Remember to read the bottom of the meniscus. Put the cap on the flask, and invert several times to make sure the solution is thoroughly mixed.

Dilution of Solutions

It is often necessary to prepare a weaker solution from a more concentrated solution by dilution. In order to accomplish this you must know the molarity of the more concentrated solution, and the volume and molarity of the solution you need to make. For example, if you needed to prepare 100 mL of a 0.15 M solution of hydrochloric acid, starting from 1.0 M HCl solution (the volume you need is dictated by how much you need and the size of the volumetric flasks available), the following procedure would be appropriate:

Since the moles of HCl will be the same in the dilute solution and the concentrated solution (you are only going to add water),

$$(\text{mol HCl})_{\text{dil}} = (\text{mol HCl})_{\text{conc}}$$

$$\text{mol HCl} = \text{molarity (mol/L)} \times \text{vol (L)}$$

$$M_{\text{dil}}V_{\text{dil}} = M_{\text{conc}}V_{\text{conc}}$$

$$0.15\ \text{M} \times 100\ \text{mL} = 1.00\ \text{M} \times V_{\text{conc}}$$

$$V_{\text{conc}} = 15\ \text{mL}$$

Therefore, 15 mL of the more concentrated solution (1.0 M) of HCl should be placed into a volumetric flask and the volume made up to 100 mL with water.

Note: **If the original solution has a concentration of more than 3 M, the acid should be added to water, rather than adding water to acid.** When water is added to a concentrated acid, the heat evolved is greater, and this may lead to splashing of the acid on the person doing the dilution.

Serial Dilutions

Often in the course of an experiment it is necessary to prepare solutions that are sequentially diluted by a factor of (for example) 10. In this case, particularly if accuracy is not critical, a good approximation can be made by taking one unit volume of the original solution and adding nine volumes of water. The initial volume used does not matter as long as you use the same volume for all measurements. This is a good method for microscale dilutions since a mark can be made on the stem of a Beral pipet. If the pipet is filled to this level every time, the same volume will be transferred.

Preparing and Using a Volumetric Pipet

To measure out 25 mL of solution by using a 25-mL pipet:

- Rinse out the pipet with two 5-mL portions of the solution.
- Holding the pipet vertically, attach a rubber bulb and squeeze the air out of the bulb. Under no circumstances should you use a pipet without a rubber bulb.

- Keeping the bulb evacuated, dip the pipet tip below the surface of the solution but not touching the bottom of the container. Release the bulb, and allow the liquid to fill the pipet until the level reaches 1 to 2 cm above the calibration mark.
- Quickly remove the bulb, and cover the tip of the pipet with your index finger. Wipe the outside of the tip with a clean piece of towel or tissue.
- With the tip of the pipet touching the wall of the source container above the level of the fluid, allow the pipet to drain until the meniscus reaches the calibration line.
- Now allow the remainder of the solution in the pipet to drain into an Erlenmeyer flask, being careful to avoid losses from splashing. When the liquid level falls below the swollen part of the pipet, touch the tip to the wall of the Erlenmeyer flask and continue draining. Do not blow out the liquid that remains in the tip.

Preparing and Using a Buret

- A buret is used for measuring varying volumes of liquids and for delivering volumes of liquids accurately. Usually, a buret is calibrated in divisions of 0.1 mL and can be read to an accuracy of 0.01 mL. To prepare a buret for use:
- Wash and rinse thoroughly with tap water.
- Rinse with two 5-mL portions of your solution.
- Clamp the buret firmly in the place where it will be used.
- Fill the buret above the zero mark with the solution; then open the stopcock and allow the solution to drain to, or just below, the zero mark, making sure the tip of the buret is filled.
- Read the bottom of the meniscus for your initial buret reading.
- After you have added the solution, do the same for the final buret reading.
- The volume of solution added is calculated by taking the difference of the two readings.

Titration

Titration is an analytical method that can be used to find the concentration of an unknown solution. Alternatively, if the concentration and volume of all solutions are known, a titration can be used to find the formula weight of a compound. The requirements for doing a titration include the following:

1. A suitable reaction to monitor, for which the full stoichiometric equation is known. Usually titrations are done on acid-base or redox reactions. The reaction must be fast and irreversible under the conditions of the titration. A reaction that comes slowly to

equilibrium would not be suitable for analysis by titration since it would take a long time to obtain readings, and accurate readings would be difficult to obtain.

2. A solution of known concentration. Make up the solution by dissolving a weighed amount of compound (solute) in a solvent to give a known amount of solution (see **Solubility Tests**), or determine the concentration of a solution already made by titrating the solution with another standardized solution.

3. A method by which the endpoint of the titration (the point where the reaction is complete and both reactants are present in equivalent amounts) may be determined. For acid-base reactions the endpoint may be determined by the use of an indicator that will change color or by the use of a pH meter (see **pH Meters** in Section 4).

4. The indicator you use will depend upon the reaction that you are monitoring. You will find a full discussion of choice of indicator in your textbook.

5. Redox reactions often have built-in endpoints, such as a color change, when the reaction reaches its equivalence point.

Common Acid-Base Indicators

Indicator	In Acid	In Base	pH Range*
Thymol blue	Red	Yellow	1.2–2.8
Bromophenol blue	Yellow	Bluish purple	3.0–4.6
Methyl orange	Orange	Yellow	3.1–4.4
Methyl red	Red	Yellow	4.2–6.3
Chlorophenol blue	Yellow	Red	4.8–6.4
Bromothymol blue	Yellow	Blue	6.0–7.6
Cresol red	Yellow	Red	7.2–8.8
Phenolphthalein	Colorless	Reddish pink	8.3–10.0

*The pH range is defined as the range over which the indicator changes from the acid color to the base color.

Reading a Meniscus

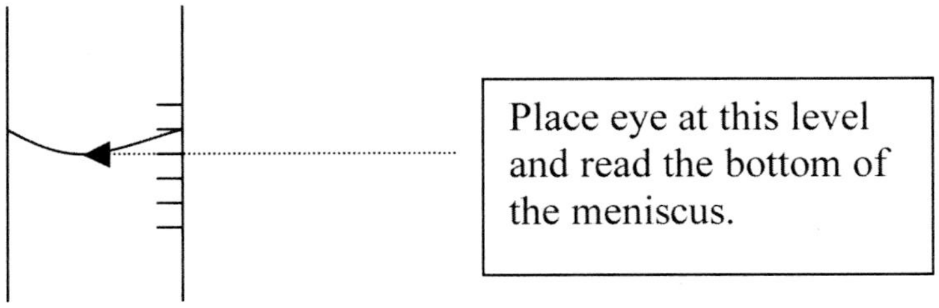

Place eye at this level and read the bottom of the meniscus.

When a liquid is placed in a container, the perimeter of the liquid is often either attracted to or repelled from the container. The most common example of this is the effect seen when water is placed in a glass container. The intermolecular forces of attraction between the water and glass are strong enough to pull the water at the edge of the container up the sides of the container, so that the water level is concave. The opposite effect is seen when mercury is placed in a glass tube: The curved interface is convex; the name of this curved interface is the meniscus. This effect becomes more pronounced as the diameter of the glass tube is decreased. It is important to read the middle part of the meniscus when using glassware with small-diameter tubes (particularly for volumetric glassware such as pipets, burets, volumetric flasks, and even measuring cylinders). To read the middle of the meniscus, make sure you place your eye at the same level as the surface of the liquid so that you get an accurate reading.

Titration Procedure

You will need to have accurate volumetric glassware to measure the volumes of your two reactants.

1. Use a volumetric pipet to measure out 25.00 mL of one reactant solution into an Erlenmeyer flask.
2. The other solution should be placed in a buret. First, wash out the buret with a small portion of the solution, and discard the solution according to the waste disposal guidelines. Then, fill the buret with solution and allow a small amount to run out so that the tip gets filled. The buret should be clamped firmly as shown.

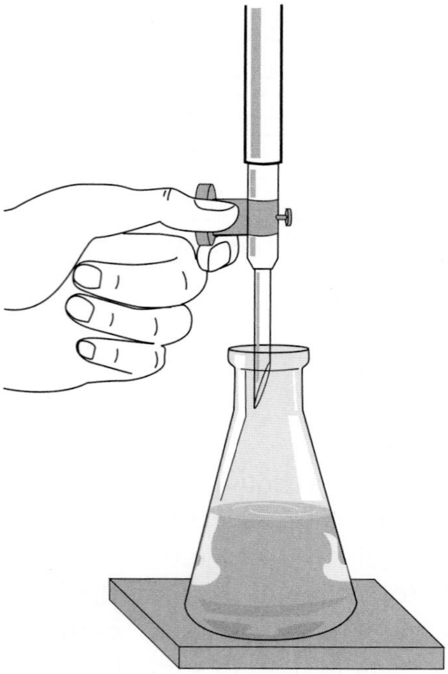

If you are right-handed, it is easiest to control the stopcock with your left hand, as shown in this picture. Your right hand can then be used to swirl the Erlenmeyer flask.

Read the buret level at the bottom of the meniscus. Add your indicator, if necessary, to the solution in the Erlenmeyer flask.

3. It is a good idea to do a rapid trial run of your titration just to get an idea of how much reagent you will need, what the endpoint looks like, and what happens when you go past the endpoint. Then, you can repeat the titration more carefully, slowing the addition to a drop-by-drop rate near the end. When you think you have reached the endpoint, record the volume added to the **nearest 0.01 mL,** and add one more drop, just to make certain that you were actually at the endpoint. Repeat the titration until you obtain two readings that agree to within 0.2 mL.
4. Calculations using the data obtained in a titration are covered fully in your text. Remember to use the average value of the two closest buret readings and the balanced equation for your calculation.

A sample calculation is given here:

If 25.00 mL of 0.09910 M H_2SO_4 takes 23.75 mL of NaOH to reach the endpoint, what is the molarity of the sodium hydroxide?

$$H_2SO_4 + 2NaOH \longrightarrow Na_2SO_4 + 2H_2O$$

$$(25.00\ \text{mL}\ H_2SO_4) \times \left(\frac{0.09910\ \text{mol}\ H_2SO_4}{1\text{L}}\right) \times \left(\frac{1\ \text{L}\ H_2SO_4}{1000\ \text{mL}\ H_2SO_4}\right)$$

$$\times \left(\frac{2\ \text{mol NaOH}}{1\ \text{mol}\ H_2SO_4}\right) \times \left(\frac{1}{23.75\ \text{mL NaOH}}\right) \times \left(\frac{1000\ \text{mL NaOH}}{1\ \text{L NaOH}}\right)$$

$$= 0.02863\ \text{mol NaOH/L}$$

A quick formula derived from this type of calculation is:

$$\frac{M_1V_1}{n_1} = \frac{M_2V_2}{n_2}$$

where M, V, and n are molarity, volume, and number of moles of each reactant, respectively.

FILTRATION

Filtration is a method used for separation of mixtures of solids and liquids. The type of filtration apparatus used depends upon whether you want the solid or the liquid from the mixture. If it is only the liquid that you want to keep, simple gravity filtration is probably the best method of separation. If it is the solid that you want, vacuum filtration may be better.

Gravity Filtration

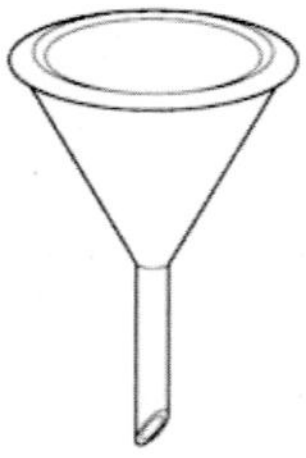

Conical funnel
(glass)

In gravity filtration, a conical funnel is fitted with a piece of folded filter paper. Wet the filter paper with the solvent you are using. (For example, if the liquid in your mixture is water, then wet the paper with water.) Pour the mixture of solid and liquid through the filter paper and funnel, and allow it to filter by gravity. The solid remains as the liquid passes through the filter paper. Do not fill the funnel above the level of the filter paper. Keep adding the mixture to the funnel as the level drops. You may find it necessary to swirl the mixture as you pour so that the solid does not get left behind in the bottom of the flask. There are different grades of filter paper, and so if your solid is very finely divided, you may have to use a different grade of filter paper. In general, the finer the filter paper, the slower the filtration. When you have filtered all the material, you may find that there is some solid left in the original flask. Pour some of the filtrate (the liquid that has been filtered) back into the flask with the solid, swirl, and refilter. Repeat this process until you have all the solid in the filter funnel. Then, wash the solid by pouring a small amount (<5 mL) of fresh cold solvent through the funnel. Repeat the washing. (Note: For washing solids, two small portions are more effective than one large portion.) The liquid (filtrate) is now ready for further use. If you also need the solid, you should allow it to dry on the filter paper before removing it to be weighed.

Vacuum Filtration

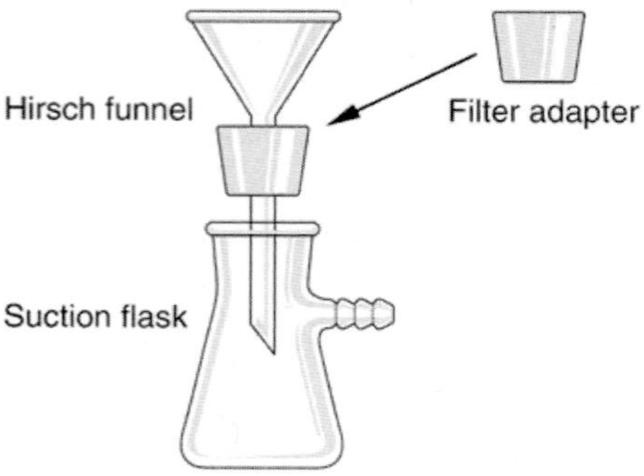

This type of filtration is usually used when it is primarily the solid that is needed. Take a Büchner funnel or a Hirsch funnel and a Büchner flask (a specialized flask with a side arm). Attach it to the aspirator (the vacuum) using a piece of thick-walled vacuum rubber tubing. Regular rubber tubing would collapse when you started the vacuum. Set up the equipment as shown, and then place a piece of filter paper over the holes of the Büchner funnel. (The paper should be flat and not folded.) Wet the filter paper with the solvent you are using, and then turn the aspirator fully on. If you only turn the water on a little bit, the water from the aspirator may get sucked back into the Büchner flask. While occasionally swirling the flask, slowly pour the mixture into the funnel and allow it to filter by the action of the vacuum. Wash out any solid that remains in the original flask using filtrate as described in the previous section. When all the material has been filtered, wash the solid as previously described. Allow the solid to dry on the filter paper, and weigh it.

Note: Before turning off the aspirator, disconnect the vacuum hose. Otherwise, water will suck back into your flask.

CHROMATOGRAPHY

Chromatography is a technique that is used to separate mixtures. It can be done in a number of ways using several different techniques, but all the techniques have the same underlying principles. One component is the stationary phase, composed of substances ranging from cellulose (in paper chromatography), to silica gel or alumina (in thin layer chromatography), to any one of a number of more complex reagents used in biological systems. The stationary phase is usually fairly polar and attracts polar substances strongly. The substances to be separated become adsorbed onto the stationary phase and will stay there unless moved by some external force.

Once the substance has been adsorbed, the stationary phase is brought into contact with the mobile phase, usually a liquid, although in some instruments the mobile phase can be gaseous. The mobile phase is drawn along the stationary phase by capillary action, and when the leading edge of the mobile phase, called the solvent front, reaches the substance, the substance is preferentially attracted to either the stationary or mobile phase, depending on the substance's polarity. Remember, like solvents dissolve like solutes. However, the attraction is seldom an all-or-nothing situation. Most substances, whether they are ionic or molecular in nature, are somewhat attracted to both phases. An equilibrium is established for the substance between the two phases, as shown in Equation 1.

$$\text{Substance-mobile phase} \rightleftharpoons \text{substance-stationary phase} \qquad (1)$$

As the solvent front moves up the paper, fresh developing solvent passes the spotted substance and new equilibria are established. At the same time, any of the substance that has dissolved in the mobile phase encounters fresh stationary phase and new equilibria are established. The overall effect of all these equilibria is that the movement of a substance depends on the nature of its relative attractions for the mobile and stationary phases. We characterize this movement in terms of a retention factor (R_f) defined in Equation 2.

$$R_f = \frac{\text{distance traveled by substance}}{\text{distance traveled by solvent front}} \qquad (2)$$

R_f values can be as high as 1.0, if the substance moves with the solvent front, and as low as 0.0, if the substance does not move at all. The values are fairly reproducible for a particular substance and solvent system, if the experimental conditions are closely controlled. One important variable is the composition of the developing solvent. If one of the solvent components is volatile, it is possible that evaporation will change the percent composition of the solvent as you develop the chromatogram. You can avoid this situation by keeping the container in which you are developing the chromatogram closed, so that the air in the container remains saturated with solvent vapor.

A sample containing two or more components can be separated, or resolved, by choosing a solvent system for which the sample components have distinctly different R_f values.

In the general chemistry lab we will probably only encounter two types of chromatography: paper and thin layer. In paper chromatography, the stationary phase is cellulose and the water adsorbed onto it. In thin layer chromatography (TLC), the stationary phase is a thin layer of silica gel or alumina coated onto a plastic or glass plate. In both cases the stationary phase is quite polar, and polar materials will adsorb very strongly. Nonpolar materials will be moved much farther by nonpolar solvents.

Thin Layer Chromatography Procedure

1. Prepare a developing chamber by placing about a 0.5-cm depth of a solvent (choose any of those provided) in a 250-mL beaker. Cover the beaker with cellophane and secure it with an elastic band or place a watch glass on top. You can also line the beaker with filter paper, which will allow the vapor and liquid to equilibrate faster. In order to have reproducible TLC results, you need to have the solvent tank saturated with the vapor of the developing solvent. Let the solvent equilibrate for 10 minutes.

2. Obtain a piece of chromatography paper or a TLC plate. If you use paper, you can fold it in half lengthwise so it will stand up in the tank. Draw a line in pencil 1 cm from the bottom of the paper.

3. Make a dilute solution of the compounds you are interested in and place a spot on the line. A drawn-out capillary tube can be used as a TLC spotter. You can then place other spots from different compounds at the same level, as in Figure 1.

Note: The spot must be small and quite dilute, otherwise the stationary phase will be overloaded and the plate will be smeared. The best way to establish this is by trial and error. If you have access to an ultraviolet (UV) visualizing lamp, you can do a preliminary check to see if you have any compound on the plate.

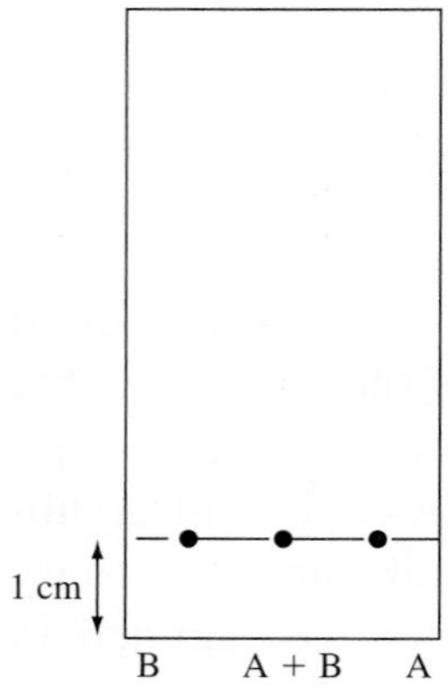

Figure 1

On the plate shown in Figure 1 two samples are spotted. Sample A might be a reaction mixture and sample B might be a starting material. The mixed spot of A and B in the middle is an excellent method for distinguishing whether sample A has any of sample B (the starting material) present.

Make sure the spots are farther from the bottom of the plate than the depth of solvent in your chamber.

4. Place the chromatography plate in the developing chamber, cover the chamber, and allow the solvent to rise up until it is about 1 cm from the top of the paper.

5. Remove the paper, and mark the level where the solvent has reached with a pencil. In your lab notebook draw a picture of the resulting separation if you can see any spots, and note the solvent system used. If nothing is visible on the plate, several methods can be used to visualize the spots. An iodine chamber will make most organic species appear as brown spots. There are a number of specialized visualizing agents that can be sprayed on the plate and heated. If the TLC plates have a fluorescent indicator in them, a UV lamp can be used; the plates will glow under the light and the spots will appear dark where they mask the indicator. If you use a UV lamp, be sure not to look directly into the light. Your lab instructor will help you decide which method is most appropriate for your needs.

   Calculate the R_f of each spot as shown in Figure 2.

$$R_f = \frac{\text{distance traveled by the spot}}{\text{distance traveled by solvent}}$$

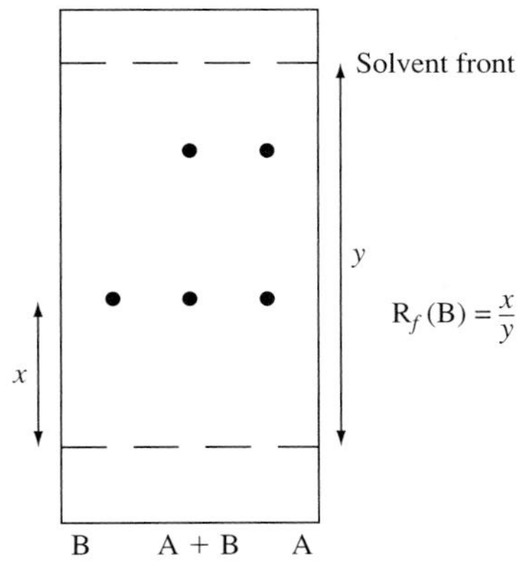

Figure 2

Note that sample A appears to contain starting material B plus a new compound. The mixed spot in the center shows that the lower compound is indeed B. If it were a different compound with the same R_f, you would see some separation of the spots at that point.

6. Repeat the process with another solvent, and note the difference in the R_f values of the spots.

7. Repeat the process until you have moved the spot to about halfway. Note which solvents move the spot a long way and those solvents that do not. Is there a correlation?

PRECIPITATION (GRAVIMETRIC ANALYSIS) FOR A SOLUTION OF A KNOWN SALT OF UNKNOWN CONCENTRATION

Suppose you have a sample of a salt in solution that you have already identified (by qualitative analysis). You may now want to find out the concentration of the salt in solution. One method is by precipitating one of the ions out of solution as an insoluble salt. This process, known as gravimetric analysis, will then allow you to calculate how much of the original salt was in solution using the formula of the salt.

For example, if you had a solution that you identified as a $BaCl_2$ solution, that is, containing barium ions and chloride ions, then you could precipitate out the barium as an insoluble salt (for example, sulfate), or you could precipitate out the chloride as an insoluble salt (for example, the silver salt). You do not need to do both, since you can calculate the mass of the other ion if you know the formula of the salt.

$$BaCl_2(aq) + Na_2SO_4(aq) \longrightarrow BaSO_4(s) + 2NaCl(aq)$$

To carry out this kind of analysis on a solution of unknown concentration:

- Measure out accurately a volume of solution using a volumetric pipet.
- Then add your precipitating agent, which must be a solution of a soluble salt containing the ion you need to precipitate your compound. For example, to precipitate barium from a solution, add a source of sulfate. This source could be sulfuric acid or a soluble salt like sodium sulfate. For either source, you must make sure that you have **excess** sulfate ions so that all the barium precipitates.
- Add the precipitating agent until no more solid seems to precipitate. Allow the mixture to sit for about five minutes, and then add a few drops more, observing to see if any more precipitation occurs. If you see more solid forming, keep repeating the steps until no more precipitation results.
- Then filter the mixture by vacuum filtration through a weighed filter paper. Wash the solid, and place it on a watch glass to dry. Allow it to dry until you get a constant weight for readings at subsequent times (at least overnight).
- Since you know the formula of the compound that has been precipitated ($BaSO_4$) and you have found the mass, you can now calculate the moles of compound that have been precipitated. From the moles of ($BaSO_4$) you know the moles of barium (Ba^{2+}) and the moles of chloride; therefore, the moles of barium chloride that were present in the original solution can be calculated and thus the concentration.

For example, if you start with 25.00 mL of a solution of $BaCl_2$ of unknown concentration, and you precipitate 0.564 g $BaSO_4$ using a solution containing excess sulfate ion, the calculation would be:

$$(0.564 \text{ g BaSO}_4) \times \left(\frac{1 \text{ mol BaSO}_4}{233.3 \text{ g BaSO}_4}\right) \times \left(\frac{1 \text{ mol BaCl}_2}{1 \text{ mol BaSO}_4}\right) \times \left(\frac{1}{25.0 \text{ mL BaCl}_2}\right)$$

$$\times \left(\frac{1000 \text{ mL}}{1 \text{ L}}\right) = 0.0967 \ \frac{\text{mol BaCl}_2}{1 \text{ L}}$$

PRECIPITATION (GRAVIMETRIC ANALYSIS) OF A SOLID UNKNOWN SALT

If you have an unknown salt, you may find that the common tests (precipitation, flame tests, or others) do not allow you to conclusively prove the identity of both ions in your solid. Then, you could use gravimetric analysis to discover the identity of the salt. However, you must know what one of the ions is before you do this.

Let's say you have identified that your compound has a chloride anion in it, but you cannot get a positive test for the cation. You suspect that it might be sodium, but it could be lithium or potassium. The following method might allow you to deduce the formula of your compound.

1. Weigh accurately about 0.5 g of your unknown, and dissolve it in water. The volume of water is not important, but you do want to make sure the solid is completely dissolved.
2. Add your precipitating agent in the form of solution. For example, to precipitate a chloride you could use a soluble salt of silver ($AgNO_3$).
3. Add the precipitating agent until no more solid seems to precipitate. Allow the mixture to sit for about five minutes, and then add a few drops more, observing to see if any more precipitation occurs. If you see more solid forming, keep repeating the steps until no more precipitate results. Then filter the mixture by vacuum filtration, through a weighed filter paper. Wash the solid, and place it on a watch glass to dry. Allow it to dry until you get a constant weight for readings at subsequent times (at least overnight). The balances we have in general chemistry labs read to 1 mg accuracy.
4. From the mass of precipitated compound and the formula (in this case AgCl), you can calculate the moles and, therefore, the mass of chloride present in the sample. From this, the mass of the cation can be calculated, and by trial and error you will be able to decide what the cation was. (Convert the mass to moles of your candidate cations. Only one possibility will be in the ratio of 1:1 or 1:2 with the moles of chloride ion.)

BOILING POINTS/MELTING POINTS

Either a commercial melting point machine, or a heating bath equipped with a thermometer, can be used to obtain the melting point or boiling point of a low melting or boiling compound. A special melting point capillary tube is used to contain the material to be tested. (A melting point capillary is a thin capillary tube with one end sealed.) The solid or liquid is placed into the melting point tube by putting the open end of the tube on the sample, causing a small amount of sample to enter the tube. The tube is then inverted, and the sealed end is tapped against the desktop so that the sample goes down to the bottom of the tube. The tube is then attached to a thermometer and placed in a water bath. As the temperature rises you will be able to monitor how the sample behaves. If the melting point or boiling point is too high for a water bath, a sand bath can be used. Specialized instruments are used for higher temperatures.

SEPARATION OF LIQUIDS

If the liquids are immiscible, that is, they do not mix, then you will see a boundary between the two, and you can use a variety of methods to separate them. For example, you can place the mixture in a measuring cylinder and pipet off the top layer. The best type of separations in this method are obtained if you use a tall, thin container so that the interface between the liquids is as small as possible. There are special pieces of equipment called separatory funnels for separating immiscible liquids, and you will probably see them in organic chemistry labs.

If the liquids are miscible, that is, they mix up completely, then you will have to find some property, chemical or physical, that will allow you to separate the liquids. For example, the physical property that you might use to separate your mixture is the boiling point. If the liquids have boiling points that are sufficiently different, you may be able to separate them by distillation, which is beyond the scope of this general chemistry lab course.

A chemical property that you could use to separate solutions is the acidity or basicity of your compounds. If you have two nonpolar (i.e., organic) compounds that are miscible with each other, one of which is acidic (i.e., it will donate its proton to a base), this difference will allow you to effect a separation. When an acid reacts with a base, it becomes ionic and thus becomes water soluble. So, if you treat your mixture with an aqueous solution of a base (e.g., NaOH), the acidic compound becomes ionic and will move into the aqueous layer. Similarly, a basic compound will move into an acidic aqueous solution. Note that this will only work if you have two nonpolar compounds since polar compounds are soluble in water anyway.

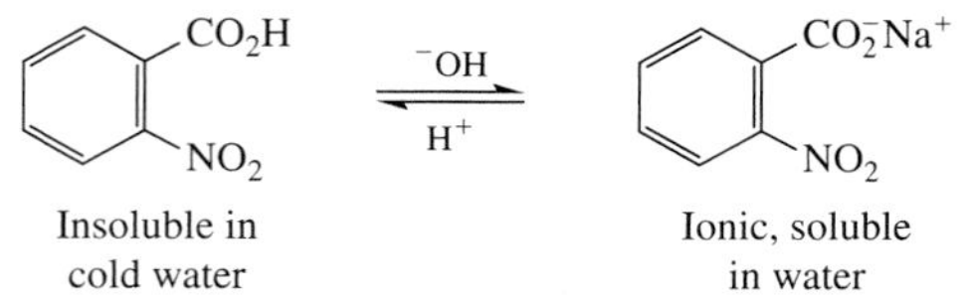

Another property used to separate compounds is polarity. The method that uses polarity as a basis for separation is chromatography. However, the use of chromatography as a method to separate large amounts (more than a few milligrams) is beyond the scope of this course.

SEPARATION OF MIXTURES OF SOLIDS

To separate a mixture of solids you will have to use some property of the components that is different for each compound. The most common property is solubility. The solubility of a solid in a liquid is governed by a number of factors. It is often possible to dissolve one component of a mixture in a solvent, leaving the other behind out of solution. The two components can then be physically separated by filtration, giving one component as a solid and the other component as a solution. Your knowledge of the type of compounds you have and their properties will help decide what solvents you should use. If all else fails, trial and error might give you some insight into a separation scheme.

Often, the results of separations give compounds that are not pure but still retain small amounts of the other components of the mixture. Purification of such solids can be effected by recrystallization.

RECRYSTALLIZATION

Recrystallization is a technique for purifying a crystalline solid. The technique involves dissolving the solid in a minimum amount of hot solvent. The solution is then allowed to cool and the crystals of pure material reprecipitate and can be filtered off. In the course of this operation the impurities stay in solution (in theory) so that the material is now free of most of the impurities. The procedure involves the following steps:

1. Find a suitable solvent. Always use small amounts (around 100 mg) for your trial recrystallizations. A test tube can be used as the container because it can be heated. If you know the nature of your compound, you can make an educated guess about what solvent to use. The ideal solvent will be one in which the compound is fairly insoluble at room temperature and soluble at the boiling point. So, for example, a nonpolar organic compound would not be soluble in water at any temperature and thus water would not be a suitable solvent. On the other hand, a nonpolar solvent might dissolve the compound even at low temperatures, and thus would not be a good recrystallization solvent, either. The best way to establish a solvent is trial and error, but you can rule out some solvents in this way.

2. Perform a test on a small scale. Add a small portion of your solvent to the test tube and record your observations. If the solid dissolves, try another solvent. When you find a solvent that does not dissolve the solid at room temperature, heat the solvent with a water bath, a steam bath, or a sand bath. **Do not use a flame unless the solvent is water**. Observe and see if the solid dissolves. If it does, cool the solution under running water while shaking, and see if crystals form. If the solid is not soluble even at high temperatures, try another solvent.
3. Cool the solution quickly and look for signs of crystallization. Smaller crystals, which are easier to handle and filter, are formed from quickly cooled solutions. If you need large crystals, allow the solution to cool slowly.
4. If no crystals have formed after about five minutes of cooling, try scratching the side of the test tube with a spatula. This will provide a rough surface that promotes crystallization. Alternatively, you could introduce a small "seed" crystal of the material on which crystallization can begin. Generally, the beginning stages of crystallization are thermodynamically unfavorable, but once the process has begun it will proceed without any further help from you.
5. If no crystals have formed even after "seeding and scratching," try boiling off some of the solvent in the hood and repeating the cooling procedure.
6. If still no crystals form, boil off all the solvent and try another solvent.
7. When you have found a good solvent and recrystallized a small batch of solid, scale up the procedure, using an Erlenmeyer flask as the container, and purify the whole amount of material. Remember, you can always get your material back by boiling off the solvent, but in that case the material will not be purified.
8. When your material has recrystallized, filter off the solid with vacuum filtration and retain the liquid; a second crop of crystals may form, but they will usually be less pure than the first crop.

ORGANIC CHEMISTRY

The methods and techniques you have learned in the lab will be a good introduction to the chemistry of organic compounds, but there are a few important aspects to organic chemistry that you should be aware of before you begin lab work.

1. Organic compounds generally are flammable. It is important to be aware of the flammability of your compounds, and **be sure that no flames are being utilized in the lab when you are dealing with organic compounds.**
2. Organic compounds may have higher toxicities than the inorganic compounds you have dealt with to date. Always be sure to treat the compounds as if they were toxic. Handle them in the hood. Do not touch, taste, or smell them. When you have identified your compound, be sure to look up the MSDS of the compound to ascertain any problems that may be associated with it.

In general the compounds you are given will not be toxic, but it is good lab practice to assume the worst.

Techniques you may need to learn when you deal with organic compounds:

1. Melting point/boiling point determinations
2. Chromatography
3. Nuclear magnetic resonance (NMR) and infrared (IR) spectroscopy

Organic Functional Group Tests

There are a number of chemical tests that can be used to determine if a specific functional group is present in a compound.

Alcohols: Ceric Nitrate Test

Many alcohols give a positive test when treated with ceric nitrate, $(NH_4)_2Ce(NO_3)_6$. The resulting complex is red.

Procedure Place about five drops of the ceric nitrate into a porcelain test plate. Add one to two drops of the unknown compound if it is a liquid. If your compound is a solid, take a few grains on the end of a spatula (about 5–10 mg) and dissolve it in a small portion (1 mL) of acetone, and then add it to the ceric nitrate. Stir with a glass rod and observe any color change.

A red color indicates that an alcohol is present. This test may also give a brown color if a phenol is present.

Aldehydes and Ketones: 2,4-Dinitrophenylhydrazine Test

Aldehydes and ketone react with 2,4-dinitrophenylhydrazine (2,4-DNP) to give 2,4-dinitrophenylhydrazones, which range in color from yellow to red.

Procedure Place a few drops of the 2,4-DNP reagent in a porcelain test plate. Add one drop of your unknown if it is a liquid. If the unknown is a solid, make a solution by adding a few grains on the end of a spatula to 5 drops of ethanol; then, add this solution to the test reagent. Stir with a glass rod and observe for a color change.

Phenols: Ferric Ion Test

Phenols form colored complexes in the presence of ferric ions. The color may be red, blue, or purple depending on the kind of phenol.

Procedure In a porcelain test plate dissolve one drop (or a few grains on the end of a spatula if a solid) of your unknown in two drops of water, ethanol, or a water ethanol mixture (you will have to test to see which system dissolves your compound the best). Add one or two drops of 2.5% aqueous ferric chloride (which is light yellow in color). Stir and observe any color change.

Section 4

Laboratory Instruments and Spectroscopy

SPECTROSCOPY

Spectroscopy can be defined as the study of the interaction of electromagnetic radiation with matter. The energy of radiation depends upon its wavelength. Different wavelengths of radiation have different interactions with matter. For example, absorption of a photon in the ultraviolet and visible region may cause an electron to move between the quantized energy levels of the atom or molecule. These are called electronic transitions. Absorption of a photon of UV light may cause an electron to move to a higher energy level. If the electron drops back down to the original level, a photon of the same wavelength will be emitted. Radiation of lower energies causes less energetic transitions between energy levels. For example, infrared (IR) radiation is of the same order of energy as the vibrational energy levels of a molecule (which are also quantized).

Nuclear Magnetic Resonance

Nuclear magnetic resonance (NMR) is a form of spectroscopy that uses the fact that certain nuclei behave like tiny spinning bar magnets. The two nuclei we will be concerned with are H-1 and C-13, both of which give rise to NMR spectra. When compounds with these nuclei are placed in a magnetic field, there are two possible orientations of the nuclei with respect to the field: a low-energy orientation in which the nuclear magnet is aligned with the field, and a high-energy orientation in which the nuclear magnet is aligned against the field. The effect of this is to split the energy levels of the nuclei. This makes it possible to cause a transition between the two energy levels by the absorption of a quantum of electromagnetic radiation of appropriate energy. The frequency of electromagnetic radiation required is in the radiofrequency range.

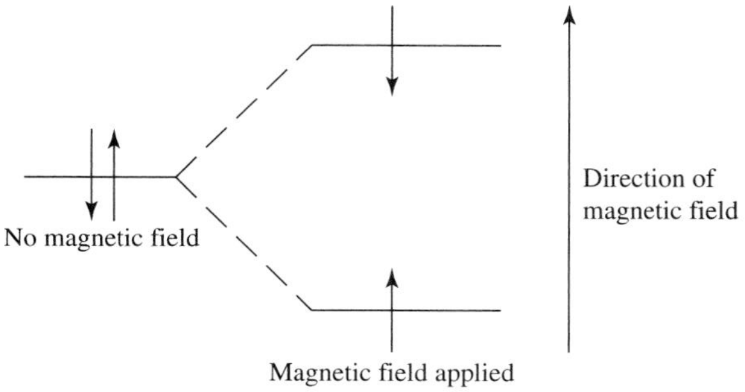

An NMR spectrometer surrounds a sample with a strong magnetic field. The sample is then irradiated with electromagnetic radiation, and it is possible to measure the energy required to jump from one level to the other. The energy of these changes depends upon the environment of the nuclei, and so from a study of the different types of absorption of energy it is possible to derive information about the nature of the compound.

The information that can be obtained from simple NMR spectra has to do with the number and type of nuclei that are in a certain compound. The simplest type of NMR spectrum is that due to C-13. (Carbon-13 is a minor isotope of carbon and is present in all naturally occurring samples of carbon compounds.) In a C-13 spectrum, each carbon atom of a given type will give rise to a signal or peak in the spectrum. The position of the peak has to do with the chemical environment of the carbon. For example, ethanol, CH_3CH_2OH, has two peaks in its C-13 spectrum because each carbon atom is in a different chemical environment.

The spectrum shown here is for cyclohexenone, a compound that has six different types of carbon atoms.

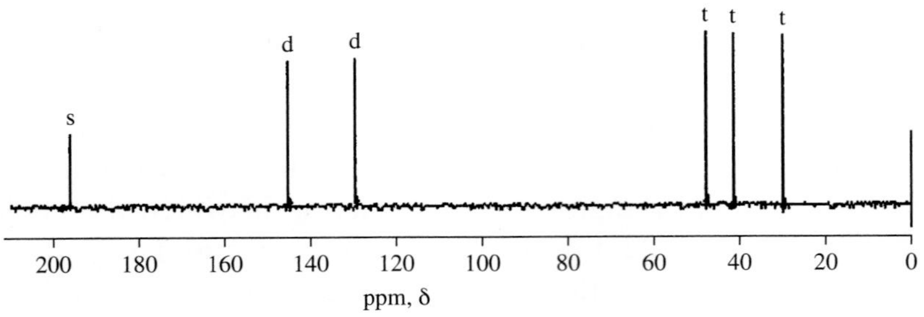

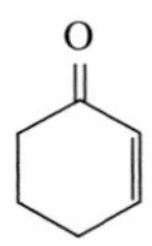

Cyclohexenone

Note that there are six signals in this spectrum and there are six types of carbon in the compound.

However, benzene, C_6H_6, has only one signal in its C-13 NMR spectrum. Since there is only one type of carbon in this molecule, all the positions in the ring are equivalent. (Draw out the Lewis structure to convince yourself that this is true.)

Proton (or H) NMR spectra appear to be more complicated since each hydrogen atom tends to give a signal that is split into several different peaks; however, for our purposes it is enough simply to compare the pattern of signals in the spectrum of your unknown with the spectra of the known compounds.

Note that the proton NMR spectrum shown here is much more complex than the C-13 NMR. However, there are five distinct clusters of signals and there are five kinds of protons in the compound that give rise to this spectrum.

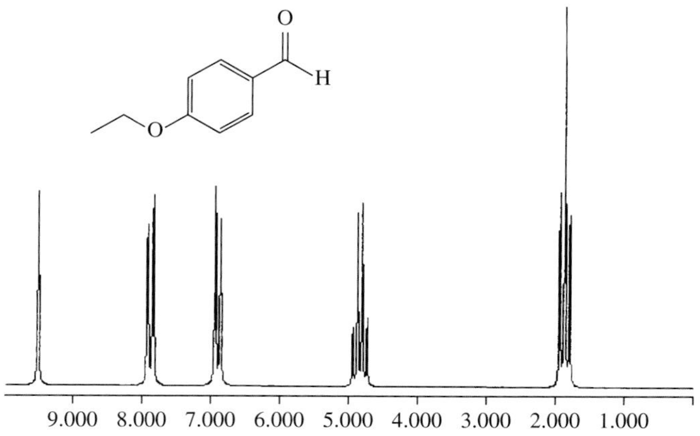

Infrared Spectroscopy

Electromagnetic radiation in the infrared range corresponds to the energy of molecular vibrations. Vibrational energy levels are quantized just as electronic energy levels are. It is possible to measure the absorption of the IR radiation as it causes the molecule to change from one vibrational energy level to another.

A typical IR spectrum is quite complicated, and a great deal of information can be obtained about the structure of the compound under investigation. In organic chemistry, the IR spectrum is very useful because particular functional groups have very specific absorptions. For example, carbonyl groups typically show absorptions in the region between 1800 and 1620 cm^{-1}. (The units on a typical IR spectrum are expressed in the reciprocal of the wavelength of radiation absorbed.)

This is a typical IR spectrum of aspirin, methyl salicylate. The most prominent feature is the strong absorption at around 1700 cm^{-1}, which corresponds to the carbonyl stretching frequency, and the broad band above 3000 cm^{-1}, which corresponds to the OH stretching frequency.

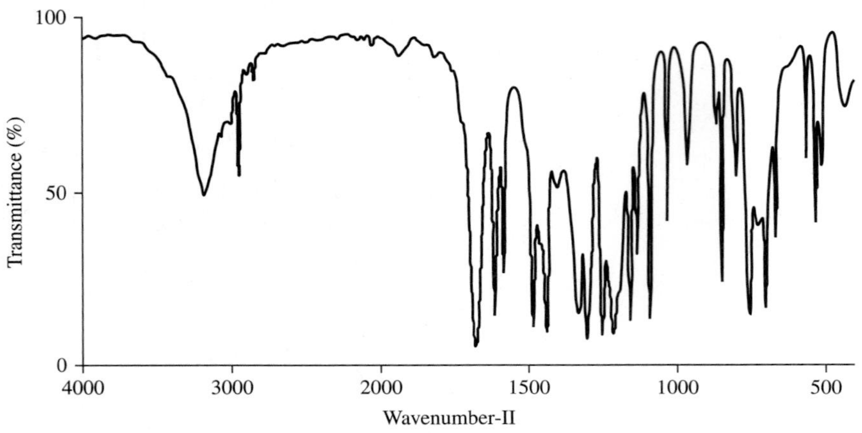

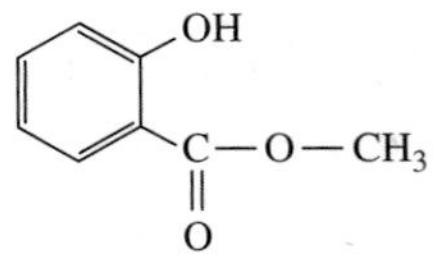

COLOR AND SPECTROSCOPY

The rainbow is an illustration of a *spectrum* in which white light is sorted into its various component wavelengths, each of which appears as a different color. Such an effect can also be produced by a prism or diffraction grating.

When various wavelengths of light enter the eye, we see different colors. The range of wavelengths visible as colors for humans is about 700 nm (nanometers, where 1 nm = 10^{-9} m) to about 380 nm. It is interesting that some animals can see colors in the wavelength region beyond violet (*ultraviolet*) or in the region beyond the red end of the spectrum (*infrared*). Humans have to rely on sophisticated instruments to make observations in these regions of the electromagnetic spectrum.

The color of light transmitted by a solution or reflected by a surface is the light that remains after some colors (or wavelengths) have been preferentially absorbed. Chemists are interested in the region in which light is absorbed because it can supply information about the structure of the molecule. We therefore need to know what wavelength is absorbed most preferentially and also the intensity of the absorption. The wavelength of absorption is related to the energy of the light being absorbed by the molecule. Thus, the color of the absorbed light provides direct information about the kind of molecular transition or the kind of molecule doing the absorbing. The intensity of absorption is related partly to the type of molecule, but it also depends on other variables.

Relationship between the Color of Light and the Wavelength of Light

Wavelength of Light Absorbed (nm)	Color of Light Absorbed
400–440	Violet
440–480	Blue
480–490	Green-blue
490–500	Blue-green
500–560	Green
560–580	Yellow-green
580–600	Yellow
600–610	Orange
610–750	Red

Remember that the color of light absorbed is not the same as the color of the solution. The color of the solution depends on the wavelengths of light **remaining** after the other light is absorbed.

The Units of Color Intensity

The meter supplies the intensity of absorption at a given wavelength. It is calibrated in the following two ways:

1. Transmittance. This is the fraction of the initial number of photons that pass through the sample without being absorbed. Usually the transmittance is expressed as a percentage $\%T$.

   $$\%T = (I_t/I_0)\,(100)$$

   where I_t is the intensity of the light after it has passed through (or been transmitted by) the solution, and I_0 is the intensity of the beam originally directed at the solution.

2. Absorbance. This is the negative logarithm of the transmittance. It is unitless. Absorbance and optical density (OD), as the biochemists call it, are the same thing. The relationship between absorbance A and percent transmittance ($\%T$) is

   $$A = 2.000 - \log(\%T)$$

The Relationship between Absorbance and Concentration

There is a direct relationship between the absorbance of a solution and the concentration of the absorbing species. This relationship is known as **Beer's law,** and it takes the form

$$A = \varepsilon c l$$

where A = absorbance
ε = constant specific to absorbing species
c = concentration of absorbing species
l = path length of cell (usually 1 cm)

pH Meter—Its Care and Use

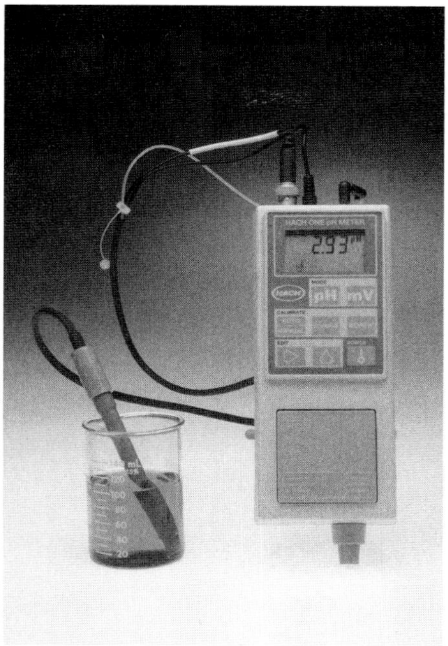

© The McGraw-Hill Companies, Inc./Charles D. Winters/Timeframe Photography.

The pH meter is actually a sensitive voltmeter that measures the potential between two electrodes: the reference electrode and the glass electrode. The glass electrode is sensitive to the pH of the solution, while the reference electrode is not. Therefore, the electrical potential that the voltmeter senses, between the reference and glass electrodes, can be transformed into a pH measurement.

Most meters have a digital display already calibrated in pH units, but, in fact, they can also be used as a voltmeter (see the following section **The pH Meter as a Voltmeter**.) The pH meters are sensitive and expensive instruments and should be treated with great care. The most sensitive part of the instrument are the electrodes, and for our purposes, a combination electrode containing both the reference and the glass electrodes in one unit will be used. When not in use, the electrode should be placed in a buffered solution so that it does not dry out. If the electrode is left out of an aqueous solution for any period of time (more than five minutes), it will have to be rehydrated for 24 hours until it stabilizes again.

When you are ready to use the pH meter, you should obtain a buffered solution, preferably pH 7. Place the combination electrode in the solution, and turn the knob on the front from standby to pH. The reading should stabilize somewhere around 7. You can use the calibration knob to make small adjustments so that the reading is exactly 7.00. If your meter will not stabilize, check to see that your solution is at room temperature. There will probably be some small fluctuations in the readings; however, if your meter is very unstable, check with your laboratory instructor.

When you have calibrated the meter, you can use it to measure the pH of your test solutions. It is good practice to wash the electrode with distilled water as you transfer it from one solution to another. When you have finished, make sure to leave the electrode sitting in the buffer solution, and switch the meter to standby.

The pH Meter as a Voltmeter

When the pH meter is used as a voltmeter, the pH electrodes are removed and replaced with wires attached to alligator clips. The clips may be attached to the electrodes of your voltaic cell, and assuming that your salt bridge is in place, the cell potential will be displayed. For normal laboratory cells, the 2-volt (V) range is probably the best setting. If the voltage appears to be negative, switch the terminal connections.

Conductivity Meter

The simplest type of conductivity meter is essentially a 9-V battery and a light bulb contained in a plastic canister. The two leads can be immersed in the solution to be tested: if the light bulb glows, the solution conducts electricity.

Section 5

Projects

PROJECT 1: DENSITY

Your team of roving troubleshooters has been called to a plastics manufacturing facility. The plastic makers are preparing a new plastic to be used in the packaging industry and need to know as much information as they can about the plastic for the licensing process. The goal for this week is to devise and carry out **two** experiments that will allow you to measure the density of the new material using the equipment available in the laboratory. Then you need to compare the two methods and advise the manufacturers which method you think is more accurate.

You will be given some pieces of the new plastic and any common equipment and chemicals you need.

> **Safety Notes**
>
> - Be sure to consult the MSDS for any compound that you work with.
> - Some of the compounds you will be dealing with may be flammable. **THERE SHOULD BE NO FLAMES IN THE LAB AT ANY TIME**. If you need to heat a solution, use a water bath heated on a hot plate.
> - Wear safety goggles and appropriate clothing at all times in the laboratory.
> - Dispose of wastes in the labeled containers. Do not pour any wastes down the drain.

Techniques You May Need to Learn or Review

- Weighing liquids and solids
- Measuring volumes

The following are some things you need to bear in mind:

1. What is density?

2. What kinds of things could you measure that would allow you to calculate density?

3. What would it mean if an object floated in a liquid without sinking or rising to the surface?

Of the two methods you devised, discuss which you think is the best and the criteria you used for defining the best method.

Before you begin, write a preliminary plan for your experimental procedure. Indicate what each person in your group will do to solve the problem and what data they will record.

PROJECT 2: INVESTIGATION OF CHEMILUMINESCENCE

We are all familiar with chemiluminescence. Fireflies are one of the more common examples of this phenomenon, but there are numbers of other naturally occurring chemiluminescent materials. There are also reactions that we can perform in the lab that produce a chemiluminescent reaction. Your company has decided to go into the production of chemiluminescent lighting. This type of lighting is ideal for use in situations where heat or electricity might be dangerous, since the production of chemiluminescent light can be contained within a plastic case and is not accompanied by the production of heat or the use of electricity.

Your task is to design a chemiluminescent light. To do this you will need to:

1. Research the chemical basis of chemiluminescence. There are numerous sites on the Internet and in basic chemistry texts.
2. In the laboratory, identify the components of the reaction that leads to the production of this light.
3. Modify the ratios of these components to extend the length of time that the light persists.
4. Investigate changes in the reaction medium that might extend the length of time that the light persists.
5. Design a case for your new lighting system that will allow it to be used on an as-needed basis without exposing the user to the chemical components.

Safety Notes

- Be sure to consult the MSDS for any compound that you work with.
- Wear safety goggles and appropriate clothing at all times in the laboratory.
- Dispose of wastes in the labeled containers. Do not pour any wastes down the drain.
- Use great care when transferring solutions of strong acids and bases.

Techniques You May Need to Learn or Review

- Use of microscale equipment (cell wells and pipets)
- Weighing solids
- Measuring liquids

The following are materials you will be given to investigate the chemiluminescent phenomena:

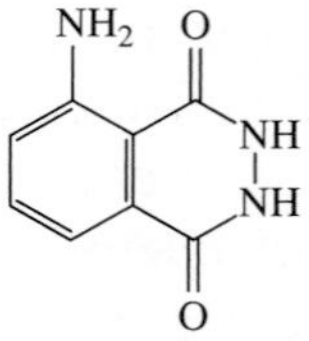

Luminol

Luminol (3-aminophthalhydrazide)
Bleach
DMSO
Air
1 M sodium hydroxide
1 M hydrochloric acid
Hydrogen peroxide

Pre-Lab Organizational Questions and Activities

1. Design a reaction protocol that will allow you to investigate the role of each of the reagents in the production of chemiluminescence.
2. What is the molecular formula of luminol?
3. List the resources that you found about the phenomenon of chemiluminescence.
4. Write a preliminary plan for your experimental procedure. Indicate what each person in your group will do to solve the problem and what data they will record.

Post-Lab Summary Questions and Activities

1. List each of the mixtures of reagents that you tried and your observations.
2. Which reagents were necessary for chemiluminescence to occur?
3. Which reagents extended the life of the chemiluminescence reaction?
4. What was the longest chemiluminescence time?
5. What do you think is causing the chemiluminescence?

PROJECT 3: CONCRETE

Your company has decided to go into the production of concrete, a building material that usually is made up of a mixture of cement, sand, and small rocks. Your group has been assigned the task of investigating the synthesis and properties of concrete. The ultimate goal for your group is to devise a recipe for the "best" concrete you can make. You will be provided with a number of materials from which concrete is usually made, in addition you will have the chemistry stockroom at your disposal should you decide to investigate some less traditional ingredients (and you should). For the purposes of your initial experiments, you will need to make a fairly small amount of concrete, and there will be several types of containers at your disposal, e.g., egg cartons, disposable paper cups, and ice cube trays.

Safety Notes

- Be sure to consult the MSDS for any compound that you work with.
- Dispose of wastes in the labeled containers. **Do not pour any wastes down the drain**.
- Use great care when transferring solutions of strong acids and bases.

Techniques You May Need to Learn or Review

- Weighing solids
- Measuring liquids
- Scaling up reactions

Pre-Lab Organizational Questions and Activities

1. Make a list of the initial mixtures you will prepare—make sure to vary the components in a systematic way so that you will be able to observe the effects of changing the amount of one component at a time.
2. How many samples of each recipe will or should you prepare?
3. How will you test the strength of your concrete?
4. Is there more than one kind of strength you should investigate?
5. Write a preliminary plan for your experimental procedure. Indicate what each person in your group will do to solve the problem and what data they will record.

Post-Lab Summary Questions and Activities

1. Make a table recording the amounts of each ingredient. Include how long you left each sample to dry.
2. Do you observe any changes (heat, appearance, etc.) as the concrete forms? What do you think is causing any changes you observe?
3. What materials did you add to your concrete samples in addition to the traditional sand, cement, and gravel chips? Were there any differences in these samples and the traditional samples?
4. How did you measure the strength of the concrete? Use at least two different methods to assess the concrete.
5. What was the recipe for the strongest concrete?

Week 2: Optional Questions and Activities

1. When you have decided which concrete is the best (using criteria of your own devising), you should scale up your recipe and make a larger sample of the concrete so that you can demonstrate its superior qualities to the rest of the class. How did you scale up your concrete recipe?
2. Did you see any difference in the small-scale reaction and the large-scale process?
3. Can you give any generalizations about what makes concrete strong and what makes it weak?

PROJECT 4: FINDING THE RELATIONSHIP BETWEEN THE VOLUME OF A GAS AND THE TEMPERATURE

This week your team has been assigned the problem of finding out what happens when you heat and cool a gas. Your company has gone into the nitrogen production business, and your boss needs to know if the temperature will affect his delivery process. The company has received an order for 10 liters of nitrogen. The boss, who is not a scientist, wants to know what the best way is to ship the gas, at high temperature or low temperature. Which way will be more cost efficient for the company? It is your job to devise experiments to find the relationship between the temperature and the volume of a gas and report back on the best way to ship an order for a particular volume of gas.

The following are extra materials available to you, besides those in your lab drawers:

"Ziplock" plastic bags
Balloons
100- or 250-mL beakers
10- or 50-mL measuring cylinders
Plastic or metal troughs
Thermometers

(You may or may not need to use all the available equipment. There is no right or wrong way to do this experiment, but some methods may be better than others.)

Techniques You May Need to Learn or Review

- Heating and cooling liquids
- Measuring volume

Pre-Lab Organizational Questions and Activities

1. You will need a closed "container" of changeable volume for your gas. What do you plan to use?
2. How will you measure the volume of the container?
3. You will need to take at least five data points at different temperatures. How will you vary the temperature and how will you measure the temperature of the gas in the container?
4. How will you accommodate the changes in volume of the gas?
5. Write a preliminary plan for your experimental procedure. Indicate what each person in your group will do to solve the problem and what data they will record.

Post-Lab Summary Questions and Activities

1. What did you use as the gas container?
2. How did you measure the volume of gas?
3. How did you vary the temperature for each reading?
4. How did you measure the temperature of the gas?
5. What relationship did you find for the volume-temperature pairs? When you report your results, you should include a graphical representation of your data.
6. At what temperature in degrees Celsius (extrapolate if necessary) would the volume of the gas reach zero? What is the significance of this temperature?
7. If you had to do this experiment again, how would you make it better?

PROJECT 5: DESIGNING A CALCIUM SUPPLEMENT

Many older people find that they have become susceptible to osteoporosis. In order to guard against this insidious ailment, doctors often recommend that people take a calcium supplement. There are a number of brands already on the market—including antacids such as Tums, and liquid supplements such as Mylanta. However, many people find these supplements "chalky," making them difficult or unpleasant to swallow.

Your task is to design a calcium supplement that could be taken as a clear liquid.

Safety Notes

- Be sure to consult the MSDS for any compound that you work with.
- Wear safety goggles and appropriate clothing at all times in the laboratory.
- Dispose of wastes in the labeled containers. Do not pour any wastes down the drain.
- Use great care when transferring solutions of strong acids and bases.

Additional Safety Note

Under no circumstances should you ever ingest any materials in the chemistry laboratory. Your criteria for making an acceptable solution should be:

- It is clear.
- It has a pH between 4 and 10.
- It does not contain any toxic materials (as indicated in the MSDS for this material).
- It has a known concentration so that people will know how much to drink to consume the required dose.

Techniques You May Need to Learn or Review

- Use of microscale cell wells and Beral pipets
- Titration
- Standardization of unknown solutions

The cheapest source of calcium is calcium carbonate, and this will be your starting point because it will not be economically feasible to use any other source of Ca^{2+} in a large-scale production.

However, before you begin to design the makeup of your solution, you will need to do some preliminary experiments to ascertain which compounds of calcium might be soluble in water.

Part I

Design experiments to investigate the solubility of calcium salts: Since it is well known that all nitrate salts are soluble, you will be provided with calcium nitrate as a source of soluble calcium. As a source of the anion you can use any sodium or potassium salt since you know that Group I salts are soluble. Be sure to make a systematic study of solubility by mixing two solutions together and observing the results. This should give you a good idea about the relative solubilities of calcium salts.

Part II

Now that you know which calcium salts are soluble and which are insoluble, you can get on with the task of making a soluble calcium solution from calcium carbonate. Your target solution will be one of the soluble salts of calcium—you can choose which one. Your task will be to take calcium carbonate, which is insoluble, and chemically alter the compound so that the calcium ions will become soluble in water rather than combining with an anion to make a precipitate.

Part I: Summary Questions and Activities

1. Draw a grid to report your observations on mixing each of the solutions.
2. For each mixture in which you observed a reaction, write the full ionic equation and the net ionic equation.
3. For a mixture in which you did not observe a reaction, write a full ionic equation and a net ionic equation.
4. What methods could you use to solubilize calcium carbonate?
5. Write a preliminary plan for your experimental procedure. Indicate what each person in your group will do to solve the problem and what data they will record.

Part II: Summary Questions and Activities

1. What chemical reactions did you perform to make the calcium carbonate soluble? Write an equation to show what happened.
2. Give a brief synopsis of the techniques you used to make the calcium carbonate soluble.
3. What is the concentration of Ca^{2+} in mol/L in your solution?
4. What is the pH of the solution?
5. How much of your solution would someone have to drink to consume the recommended daily requirement (RDA) for Ca^{2+}?

PROJECT 6: PROPERTIES OF MATTER AND SEPARATIONS

In this project, your team of troubleshooters has been called in to a landfill again. Unfortunately, the situation is more complex this time because there are a number of compounds of different types in the landfill. This time you are not limited to ionic compounds; in addition, there seems to be a mixture of compounds in the dump. Fortunately, the analytical department has been able to identify the components of the mixture. It is your task to figure out how to separate them.

To do this you will need to be able to predict, investigate, and explain the physical and chemical properties of the compounds in the mixture. From a knowledge of how the compounds behave, you will begin to be able to separate the compounds and test them to see if your predictions are correct.

You will be given a vial containing a mixture of compounds from the landfill. The identity of the compounds in the mixture is known, but the relative percentages are not. The pure compounds will be available in the lab for you to test out any separation methods before you try them on your mixture.

There are a number of preliminary steps you should take before beginning any experiments:

1. You should draw Lewis structures for each compound, making sure that you indicate the positions of all valence shell electrons. Resonance forms and formal charges should also be indicated if necessary.
2. Make ball-and-stick models where possible so that you have a good idea of the three-dimensional structure of the compounds.
3. Use a computer modeling program such as Chem3D or Spartan to develop a computer model of each compound. If possible, look at different renditions of the molecules, including electron density distributions, bond lengths, and bond angles.
4. From these structures you should try to predict the chemical and physical properties you would expect for your compounds. You will not be expected to predict absolute values for the properties but rather relative values between the different compounds you have been given.
5. Write a preliminary plan for your experimental procedure. Indicate what each person in your group will do to solve the problem and what data they will record.

The grade you receive will not depend so much on the accuracy of your predictions but rather on the explanations of your results.

Safety Notes

- Be sure to consult the MSDS for any compound that you work with.
- Some of the compounds you will be dealing with are flammable. **THERE SHOULD BE NO FLAMES IN THE LAB AT ANY TIME**. If you need to heat a solution, use a water bath heated on a hot plate.
- Dispose of wastes in the labeled containers. Do not pour any wastes down the drain.
- Use great care when transferring solutions of strong acids and bases.

Techniques You May Need to Learn or Review

- Drawing Lewis structures
- Separating liquids
- Separating solids
- Filtration
- Recrystallization

Week 1

You should be able to answer the following questions after consideration of the Lewis structures and molecular models of your compounds:

1. Which of your compounds do you think will be soluble in polar solvents? Which in nonpolar? Why? (What are the factors that govern solubility in different solvents?)

2. What are the relative melting points and/or boiling points? Many of the compounds you will be given have melting points and boiling points that are too high to measure in a general chemistry lab. Consult your instructor to find out if this is the case with your compounds. If it is, you will have to look up the melting points and boiling points of your compounds (after you have predicted their relative values).

3. Do any of your compounds have acidic or basic properties? Can you predict whether a compound will be an acid or a base from its structure?

4. Are any of your compounds electrolytes? Why?

5. What compounds do your compounds react with?

When you have all this preliminary material in hand, you can begin to prepare your separation scheme. Write a preliminary plan for your experimental procedure. Indicate what each person in your group will do to solve the problem and what data they will record.

Week 2: Summary Questions and Activities

1. What was the identity of each of your compounds?
2. What properties of these compounds did you choose to help you separate your compounds?
3. Draw a flowchart to represent your separation scheme.
4. What laboratory techniques did you need to perform to separate your compounds?
5. What was the percent recovery for each of the compounds?

PROJECT 7: ACIDS AND BASES

This month your roving team is back with the Environmental Protection Agency. The problem is that a chemical company has closed down, leaving unlabeled bottles of acids and bases in its stockroom. Before you can dispose of the bottles properly or, even better, put their contents to some good use, you will have to investigate their properties quite thoroughly. Since your company has a number of research teams, you will only have to investigate the properties of two of the unlabeled solutions. You will have access to all stock solutions (which have a molarity of 1.00 unless otherwise labeled) available in the lab. pH meters and indicators will also be provided.

Safety Notes

- Be sure to consult the MSDS for any compound that you work with.
- Wear safety goggles and appropriate clothing at all times in the laboratory.
- Dispose of wastes in the labeled containers. Do not pour any wastes down the drain.
- Use great care when transferring solutions of strong acids and bases.

Techniques You May Need to Learn or Review

- Use of pH meter
- Titration
- Titration using PASCO probeware or other pH probe/computer software
- Use of volumetric glassware
- Choosing an indicator
- Dilution of solutions
- Qualitative analysis of ions

The following are some suggestions about lines of inquiry you might want to take:

1. How will you determine if your solutions are acids or bases?
2. How will you determine the composition of your solutions?
3. How will you find the pH of your solutions? What information can you infer from the pH?
4. Do your solutions behave differently on dilution?

5. Bring into the lab some household items you consider to be acidic or basic and investigate their properties and concentrations.

You may have other ideas about how your project is to progress. This is merely a list of ideas for you to begin with.

Pre-Lab Organizational Questions and Activities

1. What is the quickest way to find out whether a compound is acidic or basic?
2. What is a more accurate way of finding out whether a compound is acidic or basic?
3. How will you find the identity of the ions in your solutions?
4. How will you find the amount of ions in your solutions?
5. Write a preliminary plan for your experimental procedure. Indicate what each person in your group will do to solve the problem and what data they will record.

Post-Lab Summary Questions and Activities

1. Describe the nature of each of your solutions in terms of concentrations and species present in the solution.
2. How did you ascertain the makeup of each solution?
3. Show the calculations for how you determined the concentration of each solution.

Week 2: Optional Questions

1. What happens to the pH of your solutions as you dilute the solution? Is there a direct relationship between the concentration and the pH?
2. What was the pH of the household items that you tested?
3. What was the concentration of the acid or base in the household items that you tested?

PROJECT 8: BUFFERS

Your team has been assigned to a medical research unit. The unit is designing and testing new pharmaceuticals; however, they have discovered that for any drug to be accepted into the body it must be administered in a buffered solution. Your task is to design and test the limits of a buffer that would be useful for this purpose.

You will be given a specific pH for which you will be asked to prepare a buffer that has a buffer capacity of 0.01 mol acid or base. You will have at your disposal the following materials:

Glacial acetic acid, benzoic acid, sodium hydrogen carbonate, sodium carbonate, sodium acetate, sodium benzoate, sodium hydrogen phosphate, sodium dihydrogen phosphate.

1 M solutions of $HCl(aq)$ and $NaOH(aq)$

Safety Notes

- Be sure to consult the MSDS for any compound that you work with.
- Wear safety goggles and appropriate clothing at all times in the laboratory.
- Dispose of wastes in the labeled containers. Do not pour any wastes down the drain.
- Use great care when transferring solutions of strong acids and bases.

Techniques You May Need to Learn or Review

- Use of pH meter
- Titration
- Titration using PASCO probeware or other pH probe/computer software
- Use of volumetric glassware
- Choosing an indicator
- Dilution of solutions
- Qualitative analysis of ions

Pre-Lab Organizational Questions and Activities

1. What is the pH that you have been asked to buffer?
2. What are the criteria for choosing the components of a buffer?
3. Which pair of reagents will you choose to prepare your buffer solution?
4. How can you calculate what the pH of a buffer solution will be?
5. What volume of buffer solution will you prepare?
6. How will you measure the buffer capacity?
7. Write a preliminary plan for your experimental procedure. Indicate what each person in your group will do to solve the problem and what data they will record.

Post-Lab Summary Questions and Activities

1. What compounds did you use to prepare your buffer?
2. Outline the calculations you used to determine the amounts of materials you used in the buffer solution.
3. Outline the procedure that you used to prepare the buffer.
4. What was the pH of your buffer? How did it compare to the calculated pH of the buffer?
5. What was the capacity of your buffer? (How much acid could you add before the pH changed markedly?)
6. What range of pH could you buffer with your proposed system?

Week 2: Optional Questions

1. Does the relative concentration of the reagents affect the pH of the buffer? (If you change one concentration and leave the other constant, does it affect the pH of the buffer?)
2. Does the absolute concentration of the reagents affect the pH of the buffer? (If you change both concentrations by the same factor, does it affect the pH of the buffer?)

PROJECT 9: WHITE POWDERS

You work in a pet store, and as part of your job you are required to keep the fish tanks clean and operational. Your instructions for doing this require that each week you add 100 g of "water conditioner" (a white solid) to the tank to maintain the correct pH and ionic strength. All goes well until one week you add the white solid to the fish tank and within 24 hours all the fish die. Clearly someone has replaced the innocuous "water conditioner" with something that is toxic to fish. It is your job to discover exactly what this material is.

You send the powder off for analysis, and the results show that there are no heavy metals and only common ions in the compound.

In the pet store you do not have much in the way of an analysis lab, but you do have standardized solutions of hydrochloric acid and sodium hydroxide and a pH meter.

Your job is to figure out what the white solid is and form a hypothesis about why it might have killed the fish.

The stockroom inventory tells you that at one time or another the pet store has had samples of sodium hydroxide, potassium hydroxide, ammonium chloride, sodium acetate, sodium hydrogen carbonate, sodium carbonate, potassium carbonate, potassium hydrogen carbonate, and potassium hydrogen phosphate.

Available materials:

5 grams white solid
1 M, 3 M, 6 M HCl
1 M, 3 M NaOH
pH meter, range of indicators, pH paper

Safety Notes

- Be sure to consult the MSDS for any compound that you work with.
- Wear safety goggles and appropriate clothing at all times in the laboratory.
- Dispose of wastes in the labeled containers. Do not pour any wastes down the drain.
- Use great care when transferring solutions of strong acids and bases.

Techniques You May Need to Learn or Review

- Use of pH meter
- Titration
- Use of volumetric glassware

- Choosing an indicator
- Dilution of solutions

Pre-Lab Organizational Questions and Activities

1. What is the goal of this experiment?
2. What do all the potential white solids have in common?
3. How can you use this property to help you analyze the compound?
4. What simple test will you run first to determine your analysis protocol?
5. Outline exactly the procedure by which you will determine the identity of the white solid.
6. How will you make sure that the results you obtain are accurate and precise?

Write a preliminary plan for your experimental procedure. Indicate what each person in your group will do to solve the problem and what data they will record.

Post-Lab Summary Questions and Activities

1. What property did you use to help you analyze the compound?
2. Describe the procedure that you used to analyze the compound.
3. What compound do you think your unknown white solid is?
4. What evidence do you have for making this claim?
5. Explain how you arrived at this answer.
6. How accurate were your measurements?
7. So, why did the fish die?

PROJECT 10: ELECTROCHEMISTRY

Your team of roving troubleshooters has been assigned to the Research and Development department of the company. The boss has discovered a stock of metals that she had forgotten about. She has had the brilliant idea that they could be used to generate electricity to power some of the lighting fixtures around the factory. She thinks this could be a cost-saving measure and has put you to work on the problem.

In this project your ultimate goal will be to construct an electrochemical cell or battery. You will be provided with a range of metals and stock solutions, and you may use any materials you consider useful after you have checked with your laboratory instructor. The final product of your project should be a more or less permanent construction, which will deliver the maximum possible voltage and light a light bulb.

For your preliminary experiments you should use the microscale equipment (cell wells, etc.). When you have devised what you think will be the optimum kind of cell, with the best electrodes, solutions, and salt bridge you can find, then you should scale up your cell and construct a container for it.

Safety Notes

- Be sure to consult the MSDS for any compound that you work with.
- Wear safety goggles and appropriate clothing at all times in the laboratory.
- Dispose of wastes in the labeled containers. Do not pour any wastes down the drain.
- Use great care when transferring solutions of strong acids and bases.

Techniques You May Need to Learn or Review

- Use of microscale cell wells
- Use of voltmeter or multimeter
- Dilution of solutions

You will be provided with a range of metals to use as electrodes and 1 M solutions of the corresponding metal ions. You may also use the graphite "lead" from a pencil to conduct electricity into your cell, rather than a strip of metal.

Pre-Lab Organizational Questions and Activities

1. What kind of reaction is needed as the basis for an electrochemical cell?

2. Which part of the reaction occurs at the anode?

3. Which part of the reaction occurs at the cathode?
4. What materials do you have at your disposal to use for each reaction part?
5. What could you make the electrodes from?
6. Write a preliminary plan for your experimental procedure. Indicate what each person in your group will do to solve the problem and what data they will record.

Post-Lab Summary Questions and Activities

1. Make a list of the different combinations of materials that you used for each electrode and the resulting voltage you observed.
2. What else did you have to include in your cell before you observed a reading on the voltmeter? Why?
3. Which combination of electrode reactions gave you the highest voltage?
4. Look up the standard potentials of each of your electrodes, and calculate the theoretical voltage for the cell(s) you have constructed.
5. How does the theoretical voltage compare with the experimental voltage you observed? What might account for any differences you may observe?
6. Should increasing the size (the amount of material but not the relative proportions) in your cell change the observed voltage?
7. What happens when you change the concentration of the solutions in your cell?
8. How might you change the concentrations to increase the overall cell potential?
9. Write a plan for your experimental procedure for next week. Indicate what each person in your group will do to solve the problem and what data they will record.

Week 2: Summary Questions

1. What happened to the voltage of your cell when you increased the size of the cell?
2. What was the highest voltage that your cell generated? What conditions did you use to produce this voltage?
3. What happens when you connect two cells together? In series? In parallel?
4. What happens when you reverse the terminals on your voltmeter?
5. Can you light a light bulb with your cell array?

PROJECT 11: IDENTIFICATION, PROPERTIES, AND SYNTHESIS OF AN UNKNOWN IONIC COMPOUND

Your group is employed by the Environmental Protection Agency as analytical chemists. An unidentified compound has been discovered in a landfill in your hometown, and your group has been given the task of investigating it. Obviously, you will want to identify the compound, but this is not the only thing you will need to do. It will be very important to the people of the area to know the properties of the compound, both chemical and physical, so that you can make predictions as to how it might behave. For example, if you know the solubility of the compound, you will be able to give some indication of whether the compound will leach out of the landfill during heavy rain. If you know what kind of reactivity the compound has, you could make some predictions on the safe disposal and the longevity of the compound. If the compound is not very reactive, it might sit in a landfill for a long time. If the compound is very reactive, it may not be as long lived, but it may react to produce something more toxic or difficult to dispose of. Therefore, it is very important that you amass as much information about the compound as you can.

Goals

1. Identify the unknown compound.
2. Discover as many chemical and physical properties of the compound as you can.
3. Devise two syntheses of the compound, and compare them for cost effectiveness, safety, and potential yield of the compound.

You will be given 5 g (no more) of the compound; you will not know the identity of the compound, nor will you be given any other information about it.

Safety Notes

- Be sure to consult the MSDS for any compound that you work with.
- All the compounds that you will work with in this project are Generally Recognized As Safe, but normal safety precautions should be observed.
- Any excess reagents, solutions, or waste materials may be disposed of by diluting the solutions and pouring them down the drain unless otherwise instructed by your laboratory teacher.

In order to help you identify your unknown compound, samples of known compounds will be available in the laboratory. Use only what you need to compare with your unknown sample in tests.

The following are some hints and ideas of possible lines of investigation for your project; however, the list is not all-inclusive and you may have other possibilities that are equally valid.

1. What solvents are your compound soluble in? What are the relative solubilities in different solvents? How will you measure solubilities? What kind of information do your results reveal about the nature of your compound? What generalities can you make about the solubility of your compound and that of other known compounds available in the lab?
2. What ions are present in your compound? How will you find out? What resources are available to you to find and learn the techniques you will need?
3. Is your compound an electrolyte? How will you find out? How does it compare to other compounds available in the lab?
4. Does your compound have acidic or basic properties? How will you find out? Will you make quantitative measurements of the acidity or basicity?
5. What compounds do your unknown react with? How did you know a reaction took place? What did you observe?
6. How will you prepare your compound? (Do not forget about stoichiometry, theoretical yield, and percent yield.) Is there more than one way to make your compound? What are the relative merits of the different methods? Do not forget safety and cost effectiveness in your deliberations.

In order to make your task feasible within a reasonable time frame, we will restrict the identity of your unknown compound to one of the following:

$NaCl$	**KCl**	**Na_2SO_4**	**$CaCl_2$**	**$MgSO_4$**
Na_2CO_3	**K_2SO_4**	**KNO_3**	**$Ca(NO_3)_2$**	**$MgCl_2$**
NH_4Cl	**$(NH_4)_2SO_4$**	**$CaCO_3$**	**$MgCO_3$**	**CH_3CO_2Na**

Samples of these compounds will be available in the lab for you to test your hypotheses and compare with your unknown.

When using a technique for the first time, use samples of known compounds from those available to practice before you use up a sample of your unknown.

Techniques You May Need to Learn or Review

- Preparing a solution (qualitative)
- Preparing a solution (quantitative)
- Measuring solution conductivity
- Analysis of ions (qualitative)
- Analysis of ions (quantitative)

- Filtration of solid
- Titration

Pre-Lab Organizational Questions and Activities

1. Outline a procedure for finding the solubility of your compound. What solvents will you use?
2. Outline a procedure for finding the quantitative solubility of your compound in water.
3. Outline a procedure for determining the conductivity of your compound. What solvent should you use for this test? If your solution conducts electricity, what does that tell you about the compound?
4. What tests will you perform to find out what anions are present in your compound?
5. What tests will you perform to find out what cations are present in your compound?
6. How will you use the known compounds that are out in the lab to help you find the identity of your unknown compounds?
7. Write a preliminary plan for your experimental procedure. Indicate what each person in your group will do next week. Remember that all tests should be run in duplicate (at least).

Week 1: Post-Lab Summary Questions and Activities

1. What is the identity of your unknown? (If you have not yet identified it, give the possibilities.)
2. Describe the experiments you carried out to determine the identity of your compound. How did each experiment lead to your identification?
3. Look up the MSDS for your compound, and record the LD_{50} and the safety precautions that should be used when handling the compound. What does an LD_{50} tell you?
4. Next week you need to make sure that your identification is correct. There are authentic samples of all the possible compounds available. You need to make a solution of your compound and a solution of an authentic sample and compare their reactivity. What kind of reactivity do you expect for your compound? (Is it acidic or basic? Will it react to give a precipitate? etc.)
5. Give five examples of reactions (neutralization, double displacement, etc.) that you can carry out next week with your compound (both your sample and the authentic sample) to investigate its reactivity and confirm its identity. Write out the expected

reactions and the products you expect to see, if any. (Remember that a negative result can still give you information.)

6. One of the techniques you will need to learn is vacuum filtration. Check out the technique in your lab manual or other resource and then give a brief description below.

7. Write a preliminary plan for your experimental procedure. Indicate what each person in your group will do to solve the problem and what data they will record.

Week 2: Post-Lab Summary Questions and Activities

1. Give the results of the five (or more) reactions that you carried out to confirm the identity of your compound. Give a brief summary of the reactivity shown by your compound. How did these reactions serve to confirm the identity of your compound?

2. In order to be sure that your identification of the compound is correct you will need to devise a method that will give a quantitative analysis of the compound. How would a quantitative identification differ from a qualitative identification?

3. Using today's results, what features of the compound could you use to give rise to a quantitative analysis? For example: Can you react your compound with something that would give an insoluble salt? Does your compound have acidic or basic properties? (Review quantitative analysis in your lab manual or other resource.)

4. Remember that quantitative analyses should be run in triplicate to give accurate results. Outline the procedure you will use to do this.

5. Outline the calculations you will use to prove the identity of your compound. (You don't need to put numbers in—just show the conversions you will do.)

6. How will the results of these calculations confirm the identity of your compound?

7. Write a preliminary plan for your experimental procedure. Indicate what each person in your group will do to solve the problem and what data they will record.

Week 3: Post-Lab Summary Questions and Activities

1. Outline three possible synthesis reactions of your compound. Give the chemical reactions.

2. Which reaction is the "best"? What criteria are you using? (Safety? Cost? Environmental impact?) Discuss each criterion.
3. Give a step-by-step outline of the experimental procedure you will use to synthesize 5 g of your compound. (Be specific.)
4. What laboratory techniques will you use during your synthesis?
5. Outline the calculations you will use to calculate the percent yield of your synthesis. (You don't need to put numbers in—just show the conversions you will do.)
6. How could you prove that you have synthesized what you intended?

PROJECT 12: HOT AND COLD

Your group has been transferred to the small-scale chemical manufacturing division of your company, and your boss (who doesn't have a good chemistry background) has been told by the technicians that they are having problems with some of the syntheses, they are doing. It turns out that when they mix some of the chemicals for their syntheses the reaction vessel sometimes gets hot and sometimes gets cold. Your boss wants to know what is going on here. Has your company discovered cold fusion? Is there some reason for this? Can the temperature changes be predicted? Your group has been given the assignment of discovering what is going on and of trying to come up with some predictions for the behavior of the systems your company is dealing with. Will this discovery make money for your company? Have you discovered an alternate energy source?

The company is interested in many different syntheses involving numerous reactions, some of which are given below. Unfortunately, the technicians did not keep good records of which reactions gave out heat and which got cold (a mistake you would never make, since you know the value of good record keeping and observations), so it is your job to investigate the behavior of these systems and classify their behavior.

Reactions Carried Out by Technicians

1. Acid-base reactions

2. Redox reactions

3. Solution of salts in water

4. Precipitation reactions

(As I said, they are not good record keepers.)

The branch of chemistry in which the energy changes associated with chemical reactions are investigated is called thermochemistry. There is a good deal of background material associated with thermochemistry in your textbook, and you are advised to review this material before you begin this project.

In this project you will be measuring the energy changes associated with some reactions. To do this you will need to construct a calorimeter, a container in which to do your reactions, which will retain all (or most) of the heat energy changes that occur during the reaction. A typical calorimeter must be a good insulator; that is, it must not allow heat transfer between the surroundings and the system under investigation. A commercial calorimeter is a very expensive item of equipment, but it is possible to construct a calorimeter in the lab that is a reasonably good insulator. The accompanying materials give advice on the construction of a calorimeter, with some practical examples of the techniques and the calculations you will need for this project.

Project Goal

To investigate the heat energy changes that occur in chemical reactions and to investigate the factors that affect heat energy changes.

In order to complete this project you will need to:

1. Construct a calorimeter.
2. Measure the heat capacity of the calorimeter.
3. Decide which reactions or processes you will investigate.
4. Decide what variations in the reactions you are going to study.

Safety Notes

- Be sure to consult the MSDS for any compound that you work with.
- All of the compounds that you will work with in this project are Generally Recognized As Safe, but normal safety precautions should be observed.
- Any excess reagents, solutions, or waste materials may be disposed of by diluting the solutions and pouring them down the drain unless otherwise instructed by your laboratory teacher.
- Use great care when transferring solutions of strong acids and bases.

Techniques You May Need to Learn or Review

- Solution dilution
- Calculation of solution concentration
- Use of Excel or other spreadsheet program to do repetitive calculations and prepare graphs.

Construction of Calorimeter

Following are two requirements for a calorimeter:

1. It should not allow heat transfer between the system under investigation and the surroundings; that is, it should be a good insulator. In other words it would be practically impossible to construct such a calorimeter in an undergraduate lab, but one of your goals will be to minimize heat loss (or gain) from the calorimeter. Typically two Styrofoam coffee cups nested inside each other, with a piece of cardboard for a lid, have been used as a calorimeter. Since you will be using your

calorimeters for a number of weeks, you will probably be able to improve on the design somewhat.

2. The calorimeter should also have a low heat capacity so that only a small amount of heat is absorbed by the calorimeter. Ideally, you would want a calorimeter with a heat capacity of zero, that is, one that would not absorb any of the heat energy. Again this goal is not practical, but it is possible to calculate the heat capacity of your calorimeter and adjust for it in your calculations.

Materials

Styrofoam cups (8 oz), cardboard lid, thermometer (try to use the same thermometer each week to prevent individual variations from thermometer to thermometer), and any other materials you feel will make the calorimeter more efficient.

Construct a calorimeter using Styrofoam cups. It will need a lid that will hold a thermometer to measure temperature changes. Try to experiment with the design so that heat loss from the calorimeter is minimized. You can check for heat loss by placing hot water in the calorimeter and monitoring the rate of temperature drop over a few minutes. Your group will need at least two calorimeters, each equipped with a thermometer.

Find the heat capacity of each calorimeter by the following method:

1. Place a measured volume of room-temperature water in your calorimeter.
2. Monitor the temperature every 30 seconds for a few minutes to make sure the temperature is stable.
3. Add an equal volume of hot water of known temperature to the calorimeter and swirl. You must use different thermometers to measure the temperature of the water in your calorimeter and the water that you are adding. The thermometer in the calorimeter is part of the setup and should not be removed. Monitor the mixture, and record the temperature every 10 seconds until the temperature starts to decrease.

Calculate the heat capacity of the calorimeter.

From the first law of thermodynamics:

Amount of heat lost by hot water	=	**amount of heat gained by cold water + heat gained by calorimeter**

Heat lost by hot water = **specific heat of water × mass of water × ΔT**

Heat gained by cold water = **specific heat of water × mass × ΔT**
and
Heat gained by calorimeter = **heat capacity (joules/°C) × ΔT**

where the specific heat of water is 4.184 joules/gram · °C J/g · °C.

The mass of water can be calculated easily since the density of water is 1.00 g/mL and you know the volume.

ΔT is the temperature change, always calculated as the final temperature (T_f) minus the initial temperature (T_i):

$$\Delta T = T_f - T_i$$

Note that ΔT will be negative for the hot water and positive for the cold water and the calorimeter, but for our purposes you can make both quantities positive. The heat change for each component in your system can be calculated by the following equations:

Calculation of Heats of Reaction

Heat change in reaction	=	**heat change of total volume of solution + heat change of calorimeter**

Heat change of solution = **total mass of solution (*m*) × specific heat (sp ht) × ΔT**

Heat absorbed by calorimeter = **heat capacity (already determined) × ΔT**

$$-\Delta H = (m \times \text{sp ht} \times \Delta T) + (\text{heat capacity} \times \Delta T)$$

Note that the sign of ΔH calculated this way must be reversed, since you are actually looking at the heat gain (or loss) by the solution, not the heat released or absorbed by the reaction.

The ΔH calculated this way must then be converted to units of kJ/mol. Therefore, it is important that you know the concentration of your solutions or the mass of reactant you have used. You can use a spreadsheet to calculate the heat change for each experiment.

Every quantity in these equations is known except the heat capacity of the calorimeter.

Measurement of Heats of Reaction

Hints and Suggestions

1. You can assume that the density and specific heats of your solutions are the same as pure water; that is, density = 1.00 g/mL and specific heat is 4.182 J/g · °C. This is

not strictly true, especially as your solutions get more concentrated. However, within the limits of our experiments this is a reasonable assumption.

2. Start out by investigating acid-base reactions, and make sure that for each reaction you calculate the heat change in kJ/mol. You will need to do these calculations in order to compare your results from different concentrations.

3. What scale will you do your reactions on? (**HINT:** For safety reasons, start with dilute solutions and proceed to more concentrated solutions in later experiments, rather than the other way around. This will give you some idea of the heat energy changes associated with the reaction and help you avoid any reactions "taking off" or becoming too energetic.)

4. What effect will making your reaction mixtures more concentrated have on the heat energy changes?

5. What effect will substituting other reactants in the reaction have? For example, does it make a difference in an acid-base reaction which acid or base you use? In a precipitation reaction, do the reactants and their concentrations affect the heat change?

Pre-Lab Organizational Questions and Activities

1. What are the goals of this project?

2. You will need to construct a calorimeter to complete this project. What are the necessary features of a calorimeter?

3. Out of what will you construct your calorimeters? What might you add to the basic materials identified in the lab manual that might improve its operation?

4. What is heat capacity? Write a simple procedure for determining the heat capacity of your calorimeter. Remember that all procedures should be done in triplicate to ensure precision.

5. Write a general equation that illustrates all acid-base reactions. Using this equation, what factors can be varied to observe their effect on ΔH?

6. How will you determine the relationship between acid-base strength (type of acid and base; strong versus weak) and ΔH? Write equations for specific reactions you will use in this determination (at least three). What variables will you need to keep constant in this determination?

7. ΔH is an extensive property. What is an extensive property? What component do the reactions in question 6 have in common that could be used in comparing the heats of reaction?

8. How will you determine the relationship between acid-base molarity and ΔH? Write specific equations for the reactions you will use in this determination (at least three). What variables will you need to keep constant in this determination?

9. The heat change value that you calculate from your experiments will be for the amount of material that you use in each reaction. How will you convert it to kJ/mol?

10. The simplest reduction-oxidation (redox) reaction to investigate is that of a metal with an acid. Write the equation for one such reaction. One reaction you want to avoid is nitric acid with iron metal. Write an equation for this reaction. Why should this be avoided?

11. Write a procedure for determining ΔH for a redox reaction and for answering the following questions (be specific in the materials and conditions you will use):

a. What is the relationship between ΔH and the type of metal used?

b. What is the relationship between ΔH and the strength (strong versus weak) of the acid used?

c. What is the relationship between ΔH and the concentration of acid used?

12. Write a procedure for determining the ΔH of solution for at least three different soluble salts. Choose the salts based on their solubilities and their availability. Look up the ΔH of solution for the salts you have chosen in the *CRC Handbook*. Compare your experimental values to those found in the handbook.

13. Using the solubility rules available in your textbook, choose two different reactions that will yield a precipitate upon mixing. Write a procedure for determining the ΔH for these precipitation reactions.

Post-Lab Summary Questions and Activities

1. Describe the makeup and construction of the calorimeter.

2. What are the qualities that a good calorimeter should have? Discuss whether your calorimeter has these qualities.

3. What are the typical values of ΔH in kJ/mol for aqueous reactions? Discuss whether you think the makeup of the calorimeter will affect the values that you obtain in your experiments.

4. What are the goals of this project? Which of these, if any, have you completed? Answer this question each week of the project.

5. Was there a difference in the temperature change recorded for each different acid-base reaction (include reactions using different strengths of acids and bases and different molarities)? Explain.

6. Was there a difference in the enthalpy change (ΔH) for each different acid-base reaction? Explain.

7. With respect to redox reactions:

   a. Is ΔH dependent on the type of metal used?

   b. On the strength of the acid used?

   c. On the concentration of acid used?

8. How were the relationships in question 7 determined?

9. How did your experimental values for solution of salts compare to values found in the *CRC Handbook*?

PROJECT 13: ANALYSIS OF COLAS

For this project your group has been transferred to the research labs of a major cola company. The company has decided to do a little industrial espionage and look at the formula for their rivals' cola products. The major ingredient that the company is interested in at the moment is the phosphate content of the colas, since many people drink cola because they think the buffering capacity of the phosphate will help to settle their upset stomach.

Your boss has just told you that he wants you to analyze as many of the rivals' cola products as possible, and he wants you to do this by using the new-fangled spectrophotometers that he paid so dearly for.

Your first goal for this project is to figure out how the Spec 20 works. There are two aspects of the Spec 20 that you will need to get straight before you begin your analyses of colas.

1. You will need to figure out the relationship between the color of a compound and the wavelength of light absorbed.

2. You will need to figure out a way to correlate the amount of light absorbed with the concentration of the species in solution (that is, figure out how you can use the Spec 20 to tell you how much of something is in solution).

Then and only then will your group be ready to analyze the cola products.

Part I: Learning about the Spectronic 20

Your first priority is to learn to use the Spec 20. Make sure you have read the instructions for use and you understand how to read both absorbance and transmittance on the instrument. Your instructor will provide you with a solution of known transmittance and absorbance. Measure both the transmittance and the absorbance, and report the values to your laboratory instructor. If your values are acceptable, then both you and the instrument have been checked out. All members of the group must measure the absorbance and transmittance and be checked out by the instructor.

Make sure that you use the same instrument for all your data.

Part II: The Relationship between Color and Wavelength of Absorbance

Your group will be given several compounds of different colors. After you have prepared solutions of appropriate concentration, it will be your task to investigate the relationship between the absorption spectrum and the color of each compound.

You might want to think about some of the following points when designing your experiments.

1. How will you make up your solutions?
2. At what wavelengths should you take readings?
3. If your solution is too strong or too weak to give accurate readings, what will you do to achieve better results?
4. Is the maximum absorption wavelength of light the same color as the compound itself? Why or why not?
5. What is the process that gives rise to the color of your compounds? Why do we see potassium permanganate as purple, for example?

Part III: The Relationship between Absorbance and Concentration

Once you have decided on the relationship between absorption and color, you should investigate the relationship between absorption of light and concentration of the solution. You need only investigate this property for one solution.

1. Which wavelength of light will you use to investigate this effect? Which wavelength do you think will give the best results?
2. There are two scales on the Spec 20, absorbance and transmittance. What is the relationship between the two? Which will you measure?
3. By what factor will you change the concentration as you monitor the absorbance and transmittance of the solution?
4. What method will you use to assess the absorbance-concentration relationship? (The most obvious method is to see if there is a linear relationship between the absorbance and the concentration. What methods do you know that might reveal such a linear relationship?)

Part IV: The Determination of Phosphoric Acid in Cola Beverages

Now that you know how to use the Spec 20 to analyze the concentration of solutions that contain colored species, you are ready to analyze your rivals' colas for phosphoric acid. The following is a general description of how the analysis is done; however, as a group you will have to devise the specific details of your experiment.

1. Unfortunately phosphoric acid itself does not absorb UV or visible light; however, it can be complexed with a reagent that produces a colored species. The reagent you

will use is an ammonium vanadomolybdate solution that will be prepared for you. You do not need to worry about the nature of this species; merely know that it is colored. You can use the results of your earlier experiments to answer these questions, to help you in your experimental design.

a. What color is the solution?

b. Which wavelengths is the solution probably absorbing?

c. So what wavelength should you take your readings at?

2. You will have to prepare a calibration curve using known concentrations of phosphate. To make up these solutions you can use KH_2PO_4 or K_2HPO_4. Remember it is just the phosphate that we are measuring. The usual range of phosphate concentration in colas is around 10^{-3} to 10^{-4} M, so make sure that you prepare a calibration curve of appropriate concentrations, but also understand that the concentrations you use will have to have absorbances that you can read on the Spec 20. Note that each point on the calibration curve should be made by mixing 5 mL of the ammonium vanadomolybdate solution with 10 mL of the phosphate solution. Make sure that you include this dilution factor in your calculations.

3. When you get to the colas, you will first need to remove the dissolved CO_2 from them (why?) by heating below the boiling point for about 20 minutes. What setup will you use to minimize evaporation? Then test your cola in the same way you did the standards, by mixing 5 mL of ammonium vanadomolybdate solution with 10 mL of the cola. You may have to dilute the cola to obtain useful readings. What kind of solution will you use for your blank?

Safety Notes

- Be sure to consult the MSDS for any compound that you work with.
- Wear safety goggles and appropriate clothing at all times in the laboratory.
- Dispose of wastes in the labeled containers. Do not pour any wastes down the drain.
- Use great care when transferring solutions of strong acids and bases.

Techniques You May Need to Learn or Review

- Quantitative preparation of solutions
- Use of volumetric glassware (pipets and burets)
- Use of the Spec 20
- Serial dilution of solutions
- Preparation of graphs (calibration curves) using a computer program such as Excel

Planning Questions and Activities

1. What are the goals of this project?
2. To be checked out on the Spec 20, you must demonstrate to your instructor that you can successfully measure the transmittance and absorbance of a solution of known concentration. Outline the procedures that should be used to measure these values.
3. How will you use compounds of different colors to determine the relationship between percent transmittance (%T) and wavelength? Or between absorbance (A) and wavelength? Outline a specific procedure to follow to accomplish this goal. Give details (describe the preparation of solutions, describe the procedure to obtain %T and A data, etc.).
4. How will you report the data from your experiments (in a table or in a graph)? Give details.
5. To determine the relationship between concentration and the quantities of %T and A, you will need to make solutions of various concentrations. You will be given a reference solution. What is its initial concentration? How will you vary the concentration of the solution? Give specific details.
6. Once the solutions are prepared, how will you determine the relationship between concentration and %T and A? At what wavelength should this be measured? Be specific.
7. Explain how you will determine the analytical wavelength for analyzing solutions for phosphate ion content. Be sure to include instructions for preparing the solution and the solution to be used as the blank.
8. What is the concentration of the phosphate solution available in the lab? How will this be used for determining the relationship between concentration and absorbance? Show sample calculations for determining the initial and final volumes necessary for diluting this solution to a desired concentration.
9. Write a procedure for preparing a calibration curve for phosphate in solution.
10. Write a procedure for determining the phosphate content in colas. Include sample preparation as well as what should be used as the blank.
11. Show a sample calculation for determining the phosphate concentration of a sample cola if the absorbance is A.

Evaluation of Lab Work

1. Which scale (%*T* or *A*) is most useful in **collecting** data from the Spec 20? Why?
2. Which scale is most useful in **reporting** data from the Spec 20? Why?
3. What was the relationship between the **color** of a solution and the **wavelength** of light absorbed?
4. Does it matter at what wavelength you measure the absorbance and transmittance? Why?
5. How will your results in your first week of experimentation help in determining the phosphate ion content of colas?
6. Summarize the results of week 2 of your investigation of colas.
7. What is the mathematical relationship between concentration and absorbance?
8. How did you find this relationship?
9. How did your experiments this week help you in your overall goal to analyze the phosphate content of colas?
10. Outline the steps you took to determine the phosphate content of colas.
11. Why did you have to treat the solution with ammonium vanadomolybdate to use this method?
12. What was the concentration of phosphate in the soft drinks you tested?
13. Are there any other methods you could use to determine phosphate content?

PROJECT 14: IDENTIFICATION, PROPERTIES, AND SYNTHESIS OF AN UNKNOWN ORGANIC COMPOUND

Your team of troubleshooters has been called to a toxic dump site. A compound has been isolated, but it is your job to identify it and find out its properties. It will be very important to the people of the area to know the properties of the compound, both chemical and physical, so that you can make predictions as to how it might behave. For example, if you know the solubility of the compound, you will be able to give some indication of whether the compound will leach out of the landfill during heavy rain. If you know what kind of reactivity the compound has, you could make some predictions on the safety disposal and the longevity of the compound. If the compound is not very reactive, it might sit in a landfill for a long time. However, if the compound is very reactive, it may not be as long lived, but it may react to produce something more toxic or difficult to dispose of. Therefore, it is very important that you amass as much information about the compound as you can.

Major Goals for Your Group

1. Identify the unknown compound.
2. Discover as many chemical and physical properties of the compound as you can.
3. Identify the functional group(s) of your compound.
4. Devise at least one reaction using the compound and materials available in the lab using reactions you have learned about in lecture or lab.

You will be given 5 grams (no more) of the compound; you will not know the identity of the compound, nor will you be given any other information about it.

When using a technique for the first time, practice using available samples of known compounds before you use up a sample of your unknown.

Known Compounds

Samples of known compounds will be available in the laboratory. Use only what you need to compare with your unknown sample in tests.

Safety Notes

- Be sure to consult the MSDS for any compound that you work with.
- Wear safety goggles and appropriate clothing at all times when in the lab.
- Some of the compounds you will be dealing with are flammable. **THERE SHOULD BE NO FLAMES IN THE LAB AT ANY TIME**. If you need to heat a solution, use a water bath heated on a hot plate.
- Dispose of wastes in the labeled containers. Do not pour any wastes down the drain.

Techniques You May Need to Learn or Review

- Melting points
- Boiling points
- Thin layer chromatography (TLC)
- Infrared spectroscopy
- Nuclear magnetic resonance spectroscopy
- Solubility testing
- Setting up a reaction
- Conductivity measurements
- Testing for acidic properties
- Heating and cooling reactions

The following are some hints and ideas of possible lines of investigation for your project. However, the list is not all-inclusive and you may have other ideas that are equally valid.

1. What physical properties of the compound can you measure? Melting points? Boiling points? R_f (see chromatography)? Will measuring some of these properties help you determine the nature of the compound?

2. What solvents are your compound soluble in? What are the relative solubilities in different solvents? How will you measure solubilities? What kind of information do your results reveal about the nature of your compound? What generalities can you make about the solubility of your compound and that of other known compounds available in the lab?

3. What functional groups are present in your compound? How will you find out? What resources are available to you to find and learn the techniques you will need?

4. Is your compound an electrolyte? How will you find out? How does it compare to other compounds available in the lab?

5. Does your compound have acidic or basic properties? How will you find out? Will you make quantitative measurements of the acidity and basicity?

6. What compounds does your unknown react with? How did you know a reaction took place? What did you observe?

7. How will you prepare your compound? (Do not forget about stoichiometry, theoretical yield, and percent yield.) Is there more than one way to make your compound? What are the relative merits of the different methods? Do not forget safety and cost effectiveness in your deliberations.

In order to make your task feasible within a reasonable time frame, we will restrict the identity of your unknown compound to one of the following:

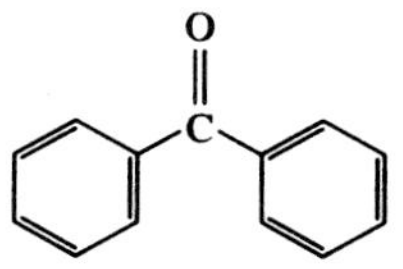

Benzophenone

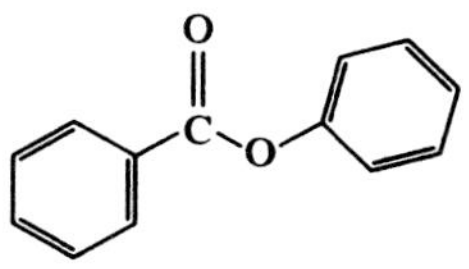

Phenyl benzoate

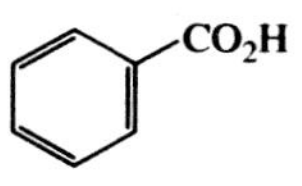

Benzoic acid

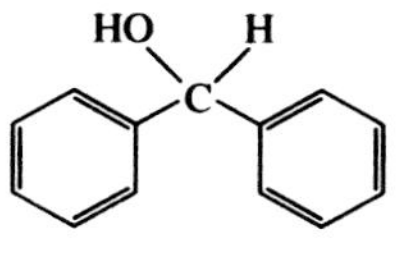

Benzhydrol

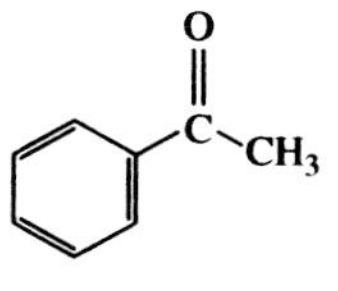

Acetophenone

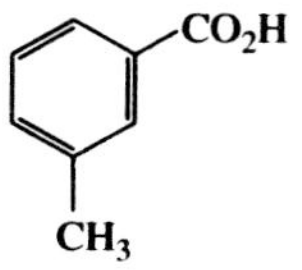

m-Toluic acid

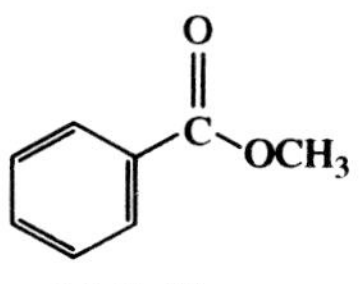

Methyl benzoate

Samples of these compounds will be available in the lab for you to test your hypotheses and compare with your unknown.

Week 1: Pre-Lab Organization Questions and Activities

1. Identify the functional groups contained in each of the unknown compounds.
2. Look up the tests for organic functional groups; which ones will you use to test your unknown?
3. Look up the melting points and/or boiling points for each of the compounds.
4. Outline a procedure for testing the solubility of your compound.
5. Write a preliminary plan for your experimental procedure. Indicate what each person in your group will do to solve the problem and what data they will record.

Week 1: Post-Lab Summary Questions and Activities

1. What functional group did your compound contain? Outline which tests you did and how you figured out which compound you have.
2. What is the LD_{50} for your compound?
3. What is the cost of your compound?
4. What is the melting point or boiling point of your compound? How close was it to the reported value in the literature? If your value is different, what might be some of the reasons?
5. What solvents are your compound soluble in? What does the solubility tell you about the nature of the bonding and intermolecular forces in your compound?
6. Look up the procedure for thin layer chromatography. Briefly outline the procedure.
7. How can you use TLC to find out if your compound is the same as an authentic sample?
8. Write a preliminary plan for your experimental procedure. Indicate what each person in your group will do to solve the problem and what data they will record.

Week 2: Questions and Activities

1. What solvent did you use to dissolve your compound in to apply it to the TLC sheet? Does it make a difference what solvent you use?
2. What solvents did you use as eluants for your TLC experiments? Does it make a difference what solvent you use?
3. What is the general effect of using a more polar solvent as an eluant?

4. Draw a picture of the TLC plates and the spots of compounds.
5. Show how you calculated the R_f of your compound.
6. Outline how you used the spectra that you either recorded or were given to help you confirm the identity of your compound.
7. Describe the reaction that you plan to do with your compound next week.
8. Write a preliminary plan for your experimental procedure. Indicate what each person in your group will do to solve the problem and what data they will record.

Week 3: Questions and Activities

1. Describe the reaction that you performed with your compound this week.
2. Write an equation to show the reaction.
3. How did you isolate your product from the reaction mixture?
4. How did you confirm that the product you isolated is in fact the product you expected?

PROJECT 15: WHAT AFFECTS THE RATE OF A REACTION?

There is a type of reaction, characterized by a sudden color change after a period of time, known as a clock reaction. This reaction could be very useful in certain circumstances—since it could be used as a timer for processes—but only if the parameters of the reaction are well known. This is where you come in. Unfortunately, the reaction is not well characterized. There is a report in the literature that indicates a mixture of the following chemical species sometimes gives rise to the phenomenon: iodide, chloride, iodate, hydrogen sulfite, and sulfate starch.

Your tasks are to:

- Identify the chemical species that take part in the clock reaction.
- Identify the parameters that affect the rate of reaction.
- Design a prepackaged timer using the reaction.

Safety Notes

- Be sure to consult the MSDS for any compound that you work with.
- Wear safety goggles and appropriate clothing at all times in the laboratory.
- Dispose of wastes in the labeled containers. Do not pour any wastes down the drain.
- Use great care when transferring solutions of strong acids and bases.

Techniques You May Need to Learn or Review

- Use of microscale cell wells and Beral pipets
- Heating and cooling reactions
- Plotting data using Excel or another graphing software

Part I: Identifying the Chemical Species

The species that might be involved in the reaction are iodide, chloride, iodate, hydrogen sulfite, and sulfate. It is also known that iodine and starch give rise to the same blue-black coloration that signifies the end of the clock reaction.

Your first task is to identify which mixture of reagents gives rise to the delayed color change that you need.

You will be provided with the following solutions:

Potassium iodide, $KI(aq)$
Sodium chloride, $NaCl(aq)$
Potassium iodate, $KIO_3(aq)$

Sodium bisulfite, $NaHSO_3(aq)$
Sodium sulfate, $Na_2SO_4(aq)$
Iodine $I_2(aq)$
Starch

Design a series of tests to see which combinations of reagents give rise to the clock reaction.

Experimental Hints

- The simplest way to do this is to perform the reaction on a microscale in a cell well plate.
- Draw a grid of the well to keep track of the mixtures you use.
- Use a different Beral plastic pipet to deliver the same volume of solution. You can make a mark with a pen about 2 inches up the stem of the pipet. Fill the pipet up to this line each time to ensure that you deliver the same volume of solution every time. It may be necessary to use more than one pipet volume to get easily seen results.
- Mix a solution of iodine and starch first so that you will be able to identify the end-point.
- Mix the same volume of each solution—in separate wells with starch to see if only one compound is responsible for the reaction.
- Then try mixtures of two of the compounds plus starch.
- If no mixture of two compounds produces the desired effect, try mixtures of three compounds plus starch.

In this way you will be able to identify which reagents are responsible for the iodine clock reaction. Your next task will be to identify the factors that affect the time delay before the clock reaction turns black.

Pre-Lab Organizational Questions and Activities

1. Design an experimental protocol that will allow you to investigate the effect of each reagent on the overall reaction. Write a preliminary plan for your experimental procedure. Indicate what each person in your group will do to solve the problem and what data they will record.

2. What do you think the starch is for?

Identifying the Reagents

1. Record your observations for each of these reactions:

$KI(aq)$ + starch
$NaCl(aq)$ + starch
$KIO_3(aq)$ + starch
$NaHSO_3(aq)$ + starch
$I_2(aq)$ + starch

What can you conclude from the results of these reactions?

2. Record your observations for each of these combinations:

$KI(aq)$ + $NaCl(aq)$ + starch
$KI(aq)$ + $KIO_3(aq)$ + starch
$KI(aq)$ + $NaHSO_3(aq)$ + starch
$NaCl(aq)$ + $KIO_3(aq)$ + starch
$NaCl(aq)$ + $NaHSO_3(aq)$ + starch
$KIO_3(aq)$ + $NaHSO_3(aq)$ + starch

3. Write an equation for the iodine clock reaction.

Part II: The Role of Each Reagent

Your next task is to see what part each reagent plays in the reaction; that is, you need to find a way to see if the amount of each reagent affects the reaction rate, and if so how it affects that rate.

There are two ways to investigate the effect of the reactant concentration on the rate of a reaction:

- Perform one reaction, and monitor the concentration of a reactant during the reaction.
- Perform many reactions, varying the initial concentrations, and monitor an aspect of that reaction, for example, initial rate or endpoint. Since the clock reaction has a built-in endpoint detector, the second method will be easier to use in this case.

The simplest way to investigate this phenomenon is to prepare a series of solutions in which the **concentration of one of the reagents is kept constant while the concentration of the other reagent is changed in a uniform way**. On the microscale, this can be accomplished by diluting a stock solution with a known volume of water.

For example, if you have a stock solution of iodate and a stock solution of hydrogen sulfite, you could start with 9 parts IO_3^- and 1 part HSO_3^- for a total volume of 10 [the units could be anything—but since you have a pipet that will deliver a certain volume, you can use the pipet volume (PV) as your unit].

If you keep the volume of HSO_3^- constant, you can dilute the iodate with water so that you keep the total volume constant. For example, you can use the following sets of reaction conditions:

HSO_3^-	IO_3^-	H_2O	Starch	Total volume
9	1	0	1 drop	10
8	1	1	1 drop	10
7	1	2	1 drop	10
6	1	3	1 drop	10
5	1	4	1 drop	10
4	1	5	1 drop	10
3	1	6	1 drop	10
2	1	7	1 drop	10
1	1	8	1 drop	10

In each of the reaction conditions, the total volume of solution stays constant; the concentration of iodate is constant, but the concentration of hydrogen sulfite varies in a uniform manner. By performing each of these reactions you should be able to investigate the effect of changing the concentration of hydrogen sulfite.

Repeating this set of reactions but varying the concentration of iodate will allow you to investigate the effect of changing the concentration of iodate.

Part III: The Effect of Temperature

How does the temperature of the reactants affect the rate of reaction? Design an experiment that will allow you to investigate this effect using at least three different temperatures.

Part IV: Designing a Chemical Clock

Now that you know what conditions affect the rate of this clock reaction, you can design a chemical clock that will allow you to measure a given time interval. Your laboratory instructor will give you the target time. It is your job to prepare a clock that will measure this time. You should design this clock so that it could be used away from the lab, so just mixing two solutions will not be appropriate.

Week 2: Pre-Lab Organizational Questions and Activities

1. What factors affect the rate of a reaction?
2. Why does each of these factors affect the rate of reaction?
3. Write a preliminary plan for your experimental procedure. Indicate what each person in your group will do to solve the problem and what data they will record.

Post-Lab Summary Questions and Activities

1. Make a table for each set of reactions showing the concentration of iodate and hydrogen sulfite and the time taken for the reaction to finish.
2. Make plots of:

   Concentration of reactant versus time
   Concentration of reactant versus 1/time
   ln concentration versus time
3. What can you conclude from each graph?
4. Write a mechanism for the reaction consistent with the rate data that you have obtained.
5. Describe how you investigated the effects of temperature.
6. Describe how you manufactured the chemical clock. How accurate was your clock?

PROJECT 16: INVESTIGATION OF KIDNEY STONES: FORMATION AND DISSOLUTION

A kidney stone is a urologic disorder that is caused by the formation of a precipitate when some soluble ions present in blood and urine react.[1] It is estimated that there were 2.7 million hospital visits and more than 600,000 emergency room visits in 2000 due to this disease. Scientists have found evidence of kidney stones in a 7000-year-old Egyptian mummy, and it is still a problem in our times, being one of the most common disorders of the urinary tract.

In order to improve the quality of life of their patients, your team of rovers has been assigned by the Kidney Stone Center of the Rocky Mountains to investigate the formation of kidney stones and to suggest ways to dissolve and prevent them.

Goals

1. Research the chemical composition of kidney stones. Numerous sites on the Internet and some chemistry textbooks are a valuable source of information.
2. Identify the major inorganic compounds present in kidney stones.
3. Prepare artificial kidney stones in a miniscale laboratory.
4. Investigate different methods of dissolving the artificial kidney stones.
5. Based on the results of the experiment, propose a strategy to prevent kidney stone formation.

Safety Notes

- Be sure to consult the MSDS for any compound that you work with.
- Wear safety goggles and appropriate clothing at all times in the laboratory.
- Dispose of wastes in the labeled containers. **Do not pour any wastes down the drain** unless instructed to do so.
- Use great care when transferring solutions of strong acids and bases.

Techniques You May Need to Learn or Review

- Solubility testing
- Use of microscale cell wells and Beral pipets

[1]http://kidney.niddk.nih.gov/kudiseases/pubs/stonesadults/index.htm (accessed November 2010).

- Measuring solids and liquids
- Dilution of solutions
- Quantitative preparation of solutions
- Calculation of solution concentration
- Precipitation reactions/recrystallization
- Use of pH meter
- Gravimetric titration
- Filtration and vacuum filtration
- Drying and weighing solids
- Qualitative and quantitative analysis of ions
- Use of Excel or other spreadsheet program to show product preparation records
- Chelating agent

Available Chemicals

$NaCl$	$Ca(NO_3)_2$	KCl
$Na_2C_2O_4$	$Mg(NO_3)_2$	NH_4NO_3
Na_2CO_3	NaH_2PO_4	Na_3PO_4
Na_2HPO_4	$CaCl_2$	$MgCl_2$
0.1 M HCl	1 M HCl	6 M HCl
0.05 M EDTA	K_2HPO_4	

Week 1

Your team should understand solubility rules and prepare four different artificial kidney stones. Remember to save your synthesized stones for dissolving them on week 2.

Week 1: Pre-Lab Organizational Questions and Activities

1. Research the compositions of kidney stones. What are the most common kinds? What are the names and formulas of the inorganic salts present in them?
2. Which combination of reagents from the list of available substances will form a precipitate? Plan a procedure to prepare and mix small amounts of different solutions for a rapid qualitative analysis. Always remember to write down all your observations. It may be helpful to create a table to input your data.
3. Outline a procedure to synthesize four of the most common types of kidney stones. For each reaction write the balanced chemical equations and net ionic equations. Do not forget to indicate which species are in aqueous solution and which are solids.
4. For each reaction show calculations of the amount of reactant required to form at least 5 g of the artificial kidney stones. Indicate which reagent you will use as the limiting reactant. Always remember to write down all your observations. It may be helpful to create a table to input your data.
5. Because your product may be suspended in solution, describe the setup on how to separate a solid from a solution.

Week 1: Post-Lab Summary Questions and Activities

1. What are the observations you gathered from the quick qualitative test?
2. Write the molecular equation and net ionic equation for each reaction indicating the aqueous and solid species. What are the resulting insoluble compounds?
3. When separating the liquids from the solids, did you have any problems? If so, how did you overcome them?

Week 2

The stones that your team synthesized last week should be dry by now. Your team's goal this week will be to devise ways to dissolve the stones prepared during week 1. Remember to prepare tables to organize the information you will gather this week.

Week 2: Pre-Lab Organizational Questions and Activities

1. What property of the artificial kidney stone needs to be measured to calculate the yield percent of your synthesis?
2. Look for a reagent that specifically interacts with the anion and cation of the artificial stones. What is the effect of the K_{sp} of kidney stones and K_a of weak acid in the dissolving process? What is the effect of the concentration of a strong acid?
3. How can the artificial kidney stones be dissolved under lab conditions? How can they be dissolved under physiological conditions? Search for a homemade remedy and the scientific basis for this treatment.
4. Propose at least three methods for dissolving the artificial kidney stones including the homemade remedy.
5. What would your team expect to be the effects of a chelating agent on the dissolving process? Try to use chemical equations to explain.
6. What would your team expect to be the effects of the pH of the solution on the dissolving process? Try to use chemical equations to explain.

Week 2: Post-Lab Summary Questions and Activities

1. Calculate the theoretical yield, experimental yield, and percent yield for each artificial kidney stone made by your group, and report this in an organized way.
2. Discuss the difference in the percent yield for each compound. Explain the factors that can possibly lead to a lower than 100% yield. Are there any factors that will result in a higher than 100% yield?
3. Discuss in the group the contributions of the K_{sp} of kidney stones, K_a of weak acid, and concentration of strong acid in the solubility of kidney stones. Give the equation for K_{sp} and K_a of the kidney stones and weak acid you analyzed.
4. Write down equilibrium equations for the solubility reactions.

Week 3

For this week your team should continue working on dissolving the artificial kidney stones and make ion tests to make sure that the stones are composed solely of the anions and cations that you expect.

Week 3: Pre-Lab Organizational Questions and Activities

1. What tests can you use to identify the cations and anions present in the artificial kidney stones?
2. How can you check if the artificial kidney stones are composed solely of the ions present in the net ionic equation?

Week 3: Post-Lab Summary Questions and Activities

1. For each of the dissolving processes you used:
   a. Show calculations, using a spreadsheet, of the theoretical and experimental amount of the dissolved salts.
   b. Discuss the possible sources of errors in the dissolving process and how to minimize, if not avoid, the errors.
2. Which of the different dissolving methods is more efficient and why? Always consider the safety and cost-effectiveness in your reflection.
3. Which of the methods that you tried can be applied to humans? What evidence do you have for making this claim? Suggest a way to apply this treatment to human kidney stones.
4. If you have to do the dissolving process again, how would you make it better?
5. Discuss with your group how you can prevent kidney stone formation using the results you obtained from the experiment.

PROJECT 17: SOAPS AND DETERGENTS

There has been another oil tanker accident. Hundreds of birds have been covered with oil, so the local environmental group has decided to help. This group, to save money, has decided to make its own soap. The problem is the only recipe they have uses lard (animal fat) to make the soap. They, being the animal lovers that they are, would like an alternative.

It is your job to develop other types of soaps and detergents for this environmental group to use on the birds. The environmental group, being all for the environment, has requested that you test the soaps, detergents, and wastes from the processes of making the soaps and detergents for environmental impact. We have included their recipe for making the soap and a well-known recipe for making detergents.

This is the second oil spill to hit the region in the past 100 years. There are horrific tales passed down about a scummy slime that was left on everything after the first oil spill was cleaned. Many suspect water contaminants were the cause of the scum. So the environmental group has asked that you also check their sources of fresh water, which come from a local pond and a well, to see if you can determine the cause and a way to prevent scum buildup.

Remember, at the end of this project you will need to make a full report on your findings. So make sure you record, either in your notebook or in printouts, all relevant data, observations, conclusions, and thoughts.

Project Goal

To make and test soaps and detergents in order to decide which one would be best for the environmental group to use in the future.
Your specific goals:

1. Test the solubility of fats, oils, soaps, and detergents.

2. Compare the desirable properties of each soap and detergent.

3. Examine the environmental impact of soaps, detergents, and respective wastewater.

4. Determine what is causing the scum after washing, and devise a solution for the problem.

5. Decide which soap or detergent is the best for this group.

Safety Notes

- Be sure to consult the MSDS for any compound that you work with.
- Wear safety goggles and appropriate clothing at all times in the laboratory.
- Dispose of wastes in the labeled containers. **DO NOT** pour any wastes down the drain, unless specifically told to do so.
- Use great care when transferring solutions of strong acids and bases.

Techniques You May Need to Learn or Review

- Use of pH meter
- Titration
- Use of indicators
- Qualitative analysis of ions
- Proper mixing of acids and bases
- Filtration
- Solubility testing
- Drying and weighing of compounds
- Vacuum filtration
- Heating reaction mixtures

To start thinking about the project, here are a few questions you might want to keep in mind:

1. What is a soap? What is a detergent?

2. What qualities will the soap need to be useful for getting the oil off the birds?

3. What properties of the soap could be possible hazards to the birds?

4. What are the causes of soap scum?

You may think of other lines of thought that help; these were just some of the questions that the environmental group had at our meeting.

Background: The Making of Soap

Soaps are made by a process called hydrolysis, when an ester is treated with water that is either acidic or basic. This process yields a carboxylic acid and an alcohol. When a basic solution is used, this process is called saponification, which means soap making.

$$R-C(=O)-O-R' + H_2O \underset{^-OH}{\overset{H^+ \text{ or}}{\rightleftharpoons}} R-C(=O)-OH + R'OH$$

Ester → Carboxylic acid + Alcohol

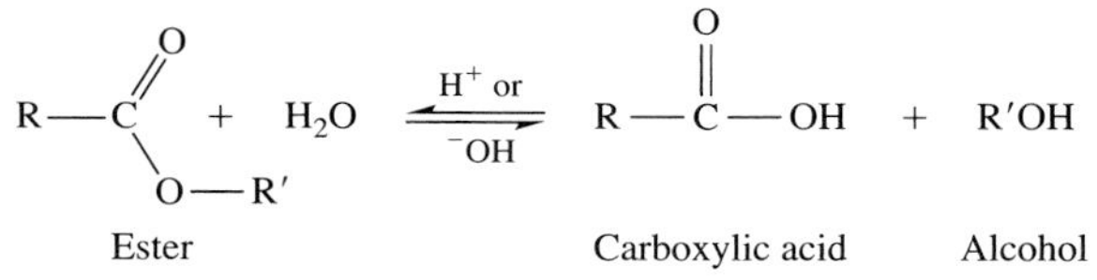

If you were to reverse the process to get the ester, the process would be called esterification. What we call fats and oils are the product of when glycerol combines with long-chain fatty acids to form an ester.

Long-chain fatty acid + Glycerol → Monoglyceride + H_2O

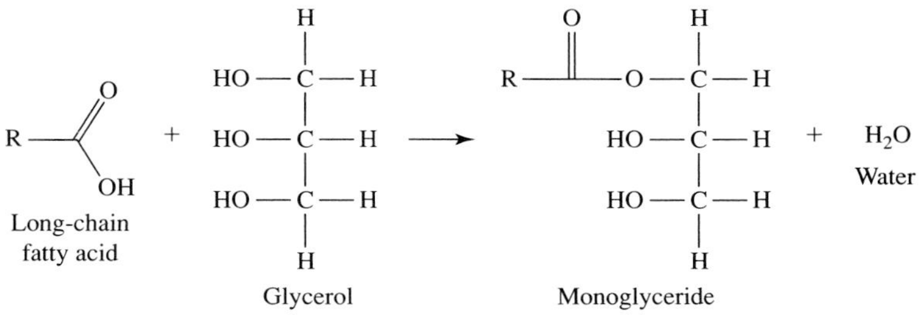

As you can see from the preceding structure, there could only be a maximum of three ester linkages onto a glycerol molecule. When there are three fatty acids attached, this structure is called a triglyceride. The way these triglycerides behave is based on the nature of the fatty acids that are attached to the glycerol. If the fatty acids are saturated, meaning that there are no double bonds, then this usually makes them solid, like the fat found in beef. Now oils, more specifically vegetable or cooking oils, which are a type of fat, have fatty acids that are called unsaturated fats. These unsaturated fats have at least one double bond in the chain. One such fat is oleic acid, $CH_3(CH_2)_7CH=CH(CH_2)_7COOH$. That would be a monounsaturated fat. When there are two or more double bonds in the carbon chain, the molecule is called a polyunsaturated fatty acid. Examples of a polyunsaturated fatty acid would be linoleic acid with two double bonds and linolenic acid with three double bonds. These types of fats are found in plants and fish.

Now, if we were to take a triglyceride and put it into a basic solution, such as a solution of sodium hydroxide, we would obtain the sodium salt of the fatty acid, and that is the soap. Soaps have a hydrophilic end, which is the carboxylic acid region, and a hydrophobic end, which is the R in the following picture but would be the chain from the fatty acid on an actual soap.

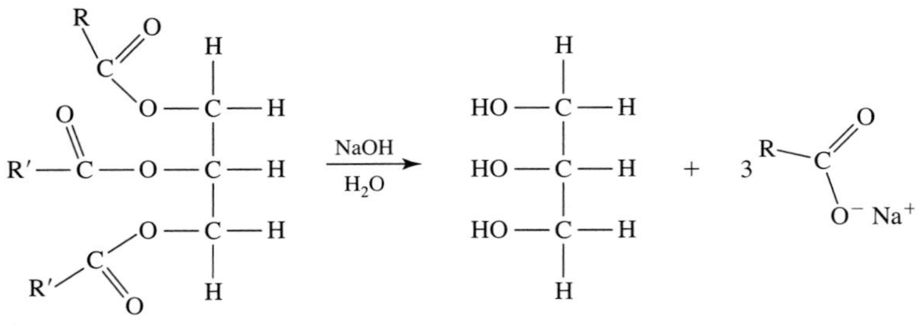

Triglyceride Glycerol Soap

Resource: General Procedure for Synthesis of Soap

1. Obtain approximately 10 mL of (10 mL of oil) oil or 10 g of (10 g of fat) fat in a 250-mL beaker. With lots of stirring, add drop by drop about 15 mL of 6 M sodium hydroxide and about 1 mL of glycerol. Stir the mixture with a glass rod.
2. Heat the solution to boiling until the solution becomes pasty. Heating the solution with a Bunsen burner is the best way to go, but be careful of material splattering out of the container. Continue heating until the mixture becomes thick and pasty.
3. Let the paste cool, and then add about 50 mL of saturated sodium chloride solution and some ice, while mixing vigorously with the glass rod. One reason for doing this is to make the slurry easy to filter.
4. Filter the soap using suction filtration, and wash it with two 5-mL portions of cold water. Separately save each filtrate for later use—you will need it.
5. Allow the soap to air dry until next week before you test it.

Background: The Making of a Detergent

Like soaps, detergents have both hydrophilic and hydrophobic ends. There are several different types of structures that can be called detergents, but the one we are looking at in this project is an anionic detergent. In fact the one that the following recipe makes is called sodium dodecyl sulfate, or simply SDS. The reason these detergents are called anionic detergents is because of the sulfur group attached to the carbon chain, and since the sulfur group has a negative charge, it gives the entire structure a negative charge.

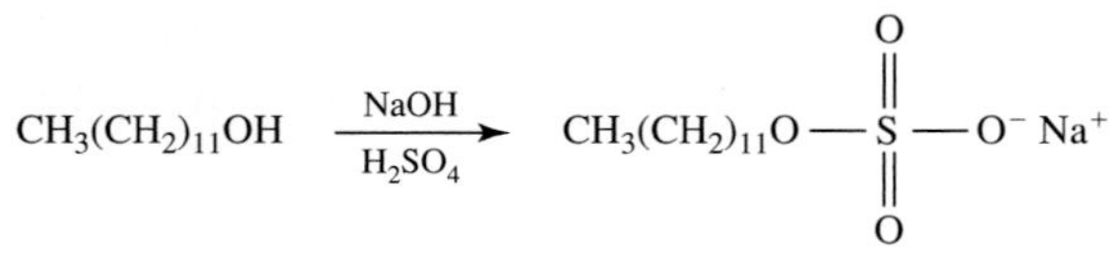

Lauryl alcohol
(dodecanol)

Sodium lauryl sulfate
[sodium dodecyl sulfate (SDS)]

Resource: General Procedure for Synthesis of a Detergent (METHOD I)

1. Place 4 mL of lauryl alcohol into a 250-mL beaker.
2. **CAREFULLY,** with stirring, add 2 mL of concentrated sulfuric acid to the lauryl alcohol.
3. Stir this mixture for another minute or so, and then let it sit for at least 10 minutes.
4. While the mixture is sitting, fill another 250-mL beaker one-third full of ice, mix into the ice about 10 g of sodium chloride, and then add water until the total volume is 75 mL.
5. In a 50-mL beaker mix 5 mL of 6 M sodium hydroxide and 10 mL of water.
6. Add 4 to 5 drops of phenolphthalein to the sodium hydroxide solution; the color could fade, but this happens in very strong basic conditions.
7. After the 10 minutes, **SLOWLY AND CAFEFULLY** add the sodium hydroxide solution to the sulfuric acid–lauryl alcohol mixture. Stir until the pink color fades.
8. Pour this solution into the salt-ice-water bath. Stir the solution to break up clumps.
9. Filter the solution using vacuum filtration. Wash the detergent twice with ice water. **Save your filtrates—you will need them later.**
10. Allow to dry.

Detergent (METHOD II)

1. In a beaker, add with **great care** 5 mL of (5 mL of lauryl) lauryl alcohol to 5 mL of (5 mL of concentrated) concentrated sulfuric acid.
2. In another beaker, add 3 drops of phenolphthalein to 10 mL of 6 M NaOH.
3. **Slowly** add the acidic solution of lauryl alcohol to the NaOH solution with constant stirring, until the pink solution just turns colorless. Keep stirring for a few minutes in an ice bath.
4. Filter the solution, and dry the precipitated sodium lauryl sulfate on the filter paper.
5. Separately save the filtrate for each batch of detergent. This will be your waste to be tested next week.

Pre-Lab Activities

1. Draw the structures of a generalized soap and of a generalized detergent, and compare the similarities and differences in the structures of soaps and detergents.
2. Describe how the structures of soaps and detergents make them good for cleaning. Use drawings to show how soaps and detergents work on a molecular level.
3. Next week each group member will make a different soap using a different starting material. Give a brief description of what each group member will be doing next week.

4. Each group will also make two batches of detergents. Outline your reaction procedure for making the detergents.
5. The environmental group is also concerned with the products you will be using to make its soaps and has asked you to test properties of the starting materials. Draw the structures of a generic oil, and predict its solubility in different solvents. Be sure to explain your predictions.
6. Design a method to test your solubility predictions for the starting materials of the soap-making process.

Week 1

This week your team will produce four soaps and two samples of detergent, as well as test the solubility of the starting materials. Remember to save the wastewater from the soap-making process for later testing. Also, remember to write down all your observations and methods used, as those will be needed in the report.

Week 1: Post-Lab Summary Questions and Activities

1. Explain the intermolecular forces at work in the solubility of the starting materials. Were your predictions correct?
2. Phenolphthalein is used in the process of making the detergents, but not in the making of the soaps. Why is this?
3. Design a plan for testing the desirable properties of soaps and detergents. **DO NOT USE YOUR OWN HANDS TO TEST AND/OR FEEL THE SOAP SOLUTIONS!** You must devise other methods of testing for this purpose.
4. How would pH affect the impact on the environment from the soaps, detergents, and wastewater, and what would a good pH be for minimizing the impact?
5. Next week you will need to analyze the wastewaters. What do you think the contaminants will be in the wastewaters, and how will you test the waste to be sure? **(Make sure you have saved your soap and detergent wastewaters separately.)**
6. Design a procedure to determine which waste is the most environmentally friendly.

Week 2

This week you will need to carry out the tests on the soaps and detergents that you designed last week. **DO NOT USE YOUR OWN HANDS TO TEST AND/OR FEEL THE SOAP SOLUTIONS!** You will also need to analyze the wastewater from the processes of making the soaps and detergents. You will need to identify the contaminants and the environmental impact of the wastewaters. You will then need to decide which soap or detergent would be the best choice for the environmental group to use. Remember to record all data, conclusions, and observations in your notebook and/or by printout.

Week 2: Post-Lab Summary Questions and Activities

1. What were the similarities and differences in the properties of the soaps and detergents you tested?
2. What contaminants did you identify in the wastewater for each soap and detergent? Explain why you would or would not consider the wastewaters dangerous to the environment.
3. Is there anything the environmental group can do to the wastewaters before disposal to make them safer?
4. Next week you will be testing the pond and well water samples. What types of contaminants may be present in these water samples that may cause soap scum?
5. Design a procedure for testing the water samples, and propose a procedure to reduce the contaminants. Keep in mind that your solution should not affect the desirable properties of the soap or detergent.
6. Also, an environmental worker spilled some of the oil on her shirt and needs help. What would be the best way for the worker to clean her shirt?

Week 3

This week your group needs to test the water samples from the well and the pond. You will also need to test the lathering abilities in these waters as well. Remember we want to minimize the environmental damage from the soaps. Your group also needs to test your suggestion for cleaning that worker's clothes. For cleaning the cloth you will be assigned one solvent or solution to use. Then at the end of the day, your group will meet with other research groups and discuss which way is the best way to clean the clothes. Remember to record all observations, conclusions, and data in your notebook or on a printout.

Week 3: Post-Lab Summary Questions and Activities

1. Did you detect any contaminants in the pond water and well water? If so, what were those contaminants?
2. How did you remove the contaminants from the pond and well water samples, and did this affect the properties of the soaps or detergents? Explain how your solution worked or why your solution did not work based on the chemistry.
3. If you were able to get the oil out of the worker's shirt, explain how. If you were unable to get the oil out of the shirt, suggest a reason as to why and a better plan that should work.
4. Using all of your previous work, which soap or detergent will you recommend to the environmental group? Explain how you made your selection by using all of the relevant information recorded during this project.

Glossary

absolute temperature The temperature scale that uses absolute zero as the lowest temperature, usually used in gas law and thermodynamic calculations. A change of 1 Kelvin is equal to a change of 1°C.

absolute zero Theoretically the lowest attainable temperature.

absorption (of electromagnetic radiation) Light is absorbed (taken up) when electrons move from a lower quantum level to a higher one.

accuracy How close a measurement is to the actual value of that quantity that is being measured. It is impossible to know whether your measurements are accurate unless you know what the actual value should be.

acid A species that donates a proton and lowers the pH.

actual yield The yield of a product you obtain from an experiment, sometimes called the experimental yield, as opposed to the theoretical yield (the amount you calculated you should get).

adsorption When one species sticks to the surface of another one. (As opposed to absorption, i.e., being taken into the body of the other species.)

air bath A reaction may be heated by placing it in an evaporating dish on top of an empty beaker. The beaker can then be heated with a Bunsen burner, and the resulting temperatures will be higher than heating over a water bath. Caution: This method should only be used when you have a reaction or process that is not affected by high temperatures.

amphoteric Refers to a species that has both acidic and basic properties.

analyte The species being analyzed.

anhydrous Without water.

anion A negatively charged ion.

anode The electrode at which oxidation takes place.

aqueous In water. An aqueous solution has water as the solvent.

arrhenius acid A substance that increases the hydrogen ion concentration in water.

arrhenius base A substance that increases the hydroxide ion concentration in water.

balance An instrument used to find the mass of objects.

balanced equation A chemical equation that has the same number of atoms of each type on both sides of the equation.

base A species that raises the pH of an aqueous solution.

beaker A straight-walled container.

Beer's Law The absorbance (A) of a solution is directly proportional to the concentration of the solution and the path length through which the light passes.

boiling point The temperature at which the vapor pressure of a liquid is equal to the external pressure. The boiling point at one atmosphere pressure is the normal boiling point.

Bronsted-Lowry acid A substance that acts as a proton donor.

Bronsted-Lowry base A substance that acts as a proton acceptor.

buffer A solution that resists changes in pH.

Bunsen burner A gas heater used in the lab.

buret A device for delivering variable accurate amounts of liquid.

cathode The electrode at which reduction occurs.

cation A positively charged ion.

centrifuge A lab instrument for separating liquids and solids by centrifugal force.

chemical properties Properties of a species that cannot be studied without changing that species into a different chemical compound.

chromatography A method of separation of species that relies on differing polarities of the compounds.

combustion The reaction that occurs when a species is burned in air.

compound A substance composed of two or more elements chemically combined in fixed proportions.

condensation To change from a gas to a liquid. Also the name given to reactions (other than acid-base reactions) that produce water as one of the reactants.

cooling bath A bath that is used to cool a reaction or system under investigation. Usually comprised of a water and ice mixture and reaching temperatures as low as 0°C. For temperatures lower than that, a mixture of ice, water, and salt solutions are used. The temperature of such baths will decrease to about −10°C. (Why?) To reach even lower temperatures, mixtures of dry ice and organic solvents can be used.

crucible A porcelain dish used to heat species to very high temperatures.

decant To pour a liquid carefully off the top of a solid or another liquid layer so that the bottom layer is not disturbed.

decantate The liquid that is decanted off.

decomposition A reaction that leads to the breakdown of a compound into simpler substances.

density The mass (m) of a substance divided by its volume (V): $D = m/V$.

desiccant A species that will remove water. Often used to dry compounds or reaction mixtures. For example, calcium chloride will absorb moisture from its surroundings, thus leaving the surroundings (usually a reaction mixture) drier.

dissolve To make a homogeneous mixture of two or more components.

distill To separate or purify a mixture of liquids on the basis of differences in boiling points.

dropper A piece of equipment for delivering drops of liquid.

effervescence The evolution of gas.

Erlenmeyer flask A glass container with sloping walls.

evaporating dish A porcelain dish used to evaporate liquids from a solution.

extensive properties Properties that depend on the amount of material being studied. For example, mass is an extensive property.

filter To separate a solid and a liquid by passing the mixture through a funnel lined with filter paper.

filtrate The liquid that remains after the solid has been removed by filtration.

forceps A tool used to pick up small objects.

funnel A tool used for filtering (with a filtering aid like filter paper) or for directing the flow of liquid into a container. There are several types.

graduated cylinder A device used for measuring volumes of liquid.

heating bath A bath used to heat reactions or experiments. There are several ways to set up a heating bath, including a water bath, a sand bath, and an air bath.

heterogeneous mixture A mixture in which the individual components remain physically separated and can often be seen as separate components. Sampling the mixture at different places would give different compositions.

homogeneous mixture A mixture in which the composition is the same throughout the sample.

hydrate A compound that incorporates a fixed number of water molecules within its structure.

hydrolysis A reaction in which water is one of the reactants.

ignition tube A large test tube that can be used for reactions or for heating things strongly.

indicator A substance that changes color according to the pH. Often used to give the endpoint of a titration. The indicator must be chosen carefully. Otherwise, misleading results may be obtained.

intensive properties Properties that are independent of the amount of matter, for example, density.

Lewis acid An electron pair acceptor. Note: Do not become confused with the definition of a Lewis acid and a Bronsted-Lowry acid. In many instances a proton donor is also an electron pair acceptor. The Lewis definition of an acid is a more encompassing definition that includes Bronsted-Lowry acids within its guidelines.

Lewis base An electron pair donor. Note: Do not become confused with the definition of a Lewis base and a Bronsted-Lowry base. In many instances a proton acceptor is also an electron pair donor. The Lewis definition of a base is a more encompassing definition that includes Bronsted-Lowry bases within its guidelines.

melting point The temperature at which the solid and liquid phases of a substance are in equilibrium. Melting point is a misnomer; because most substances melt over a range of a few degrees of temperature. In general, the shorter its melting point range, the purer the substance.

meniscus The shape of the top of a column of liquid. This shape depends on the nature of the liquid and the material the container is made of. For example, the meniscus for a column of water contained in glass is "U" shaped, and the meniscus for mercury in glass is "n" shaped. Regardless of the shape, the level of liquid is always read from the center of the meniscus, with the eye at the same height as the liquid level.

microscale A system of equipment and techniques that allows reactions and observations to be performed on a very small scale. The benefits are decreased wastes and a reduced chance of accidents.

molality (m) A concentration term used in colligative property calculations, defined as moles of solute per kilogram of solvent (mol/kg).

molarity (M) The most common concentration term, defined as moles of solute per liter of solution (mol/L).

neutralization The reaction of an acid with a base to produce salt and water.

oxidation The loss of electrons.

percent yield The yield of a reaction expressed as the actual yield divided by the theoretical yield multiplied by 100.

pH The negative log of the hydrogen ion concentration ($pH = -\log [H^+]$).

physical properties Those properties that can be studied without changing the chemical makeup of the substance.

pipestem triangle A support for a crucible that allows it to be heated strongly.

pipet A piece of equipment for delivering reagents drop by drop. Can be made of glass with a rubber bulb (Pasteur pipet) or of plastic with a built-in bulb (Beral pipet).

precipitate A solid that comes out of solution as a result of a reaction.

precipitating agent A reagent that will bring about precipitation from a given solution.

precision The agreement between two or more measurements of the same entity.

qualitative Refers to what types of substances are present.

quantitative Concerns how much substance is present.

recrystallization The process by which crystalline solids can be purified. Ideally, the solid is dissolved in a minimum of hot solvent. The solution is then allowed to cool, and crystals of the pure solid will form which can then be filtered. In practice, trial and error on a number of solvents is usually needed to find a suitable solvent for recrystallization.

reduction The gain of electrons.

***R_f* value** In chromatography, the distance moved by the spot of substance divided by the distance moved by the solvent front. It can be quoted as a property of the substance in a given solvent system and stationary phase. In practice, chromatography systems vary so much that the Rf alone cannot be used to identify a compound.

saturated solution A solution in which no more solute can be dissolved.

solubility The amount of solute that can be dissolved in a given amount of solvent at a given temperature. Reported as grams per liter (g/L) or moles per liter (mol/L).

solubility rules A generalized set of rules that describe the solubility of most ionic substances in water.

solute The minor component of a solution.

solution A homogeneous mixture of two or more components.

solvent The major component of a solution.

stoichiometry The mass relationships between the reactants and products in chemical reactions.

supernatant liquid The liquid left after a precipitate is formed.

tare The process by which the weight of a container is accounted for when weighing a substance. The balances in our labs will automatically tare when the menu bar is pressed after an empty container is placed on the balance pan.

test tube Glass or Pyrex cylindrical container that can be used for small-scale experiments.

theoretical yield The maximum amount of product that could be produced from a reaction, calculated from the stoichiometric relationships in the reaction. In practice the theoretical yield is never attained. For example, a reaction may not go to completion, or mechanical losses in handling the product always lead to an actual yield that is less than the theoretical.

titration An analytical technique that allows you to find an unknown concentration or formula weight.

tongs A tool used to transfer hot objects from place to place.

universal clamp A clamp that at one end will attach to the stand on the bench top. The other end can be used to secure equipment such as test tubes or funnels.

volumetric equipment Glassware that will hold or deliver a very accurately measured volume of liquid.

watch glass A shallow glass bowl used to allow crystals to dry.

water bath Usually a beaker of water heated over a Bunsen burner. Used to heat reactions and experiments up to 100°C.

Index